Fábio Leonardo Tomas

FábioInfluence of forest biodiversity on the occurrence of insect pests

Fábio Leonardo Tomas

FábioInfluence of forest biodiversity on the occurrence of insect pests

Tree diversity as a tool for pest control in tomato growing

ScienciaScripts

Imprint
Any brand names and product names mentioned in this book are subject to trademark, brand or patent protection and are trademarks or registered trademarks of their respective holders. The use of brand names, product names, common names, trade names, product descriptions etc. even without a particular marking in this work is in no way to be construed to mean that such names may be regarded as unrestricted in respect of trademark and brand protection legislation and could thus be used by anyone.

Cover image: www.ingimage.com

This book is a translation from the original published under ISBN 978-3-330-99729-5.

Publisher:
Sciencia Scripts
is a trademark of
Dodo Books Indian Ocean Ltd. and OmniScriptum S.R.L publishing group

120 High Road, East Finchley, London, N2 9ED, United Kingdom
Str. Armeneasca 28/1, office 1, Chisinau MD-2012, Republic of Moldova, Europe
Managing Directors: Ieva Konstantinova, Victoria Ursu
info@omniscriptum.com

Printed at: see last page
ISBN: 978-620-8-61269-6

DEDICATORY

The memory of Flávio José Thomaz, father and best friend.

To our guardian angel, Beatriz Tomaz Rubel

To all the children in the Thomaz and Alecrim family.

ACKNOWLEDGEMENTS

To God:

To my wife Daniela, for her strength throughout this journey we have travelled together.

To Mum Darticléa and Aunt Geralda, for their years of dedication and faith.

To my children, João, Isadora and Flávia Helena, for their strength.

To the whole family, Daniele, Angela, Ana Carolina, Zelma, Alessandro and Américo, for all the beautiful life together.

To Professor and supervisor Fábio Poggiani, for his trust

To my tutors in work and in life, Professors Paulo Yoshio Kageyama and Hasime Tokeshi, God bless you.

To the teachers and friends who supported this project, Flávio Gandara and Miguel Cooper.

To the farmers of the "Luiz David de Macedo" Rural Settlement, Zaqueu, Toninho, José Pinheiro and Joaquim, for their example of walking the path and striving to overcome challenges, in this project and in everyday life.

To the farmers of the Encapoeirado neighbourhood and Apiaí, for their valuable partnership

To the whole team at LARGEA and PPDARAF - ESALQ, João Dagoberto, Pedro Salústio, João Portella, Marcos "Zidane", Andréa, Elza, Talita, for their daily support, patience and companionship.

To the INCRA technicians and the Rural Settlement Prod. Rural Settlement - PDS "Luiz David de Macedo", Luiz "Timbalada", Osvaldo, Rosa and so many others who have shown their willingness to build our new society at all times.

To the technicians who helped with this production, Daniel Tomazeto, Maíra Cotrim, Renata Morelli, Bruno Brasil, Renato "Solei", Leandro, Kátia Pisciotta, Maurício Marinho, who lovingly helped solve some of the challenges of this work.

The Antro sorority and their entire family for their unconditional strength and home.

To all the administrative staff of the Postgraduate Programme in Forest Resources, Catarina Germuts and colleagues for their support.

The team at the Forestry Foundation and PETAR, Donizetti Barboza, Antônio Modesto, Tadeu Gonçalvez, Nilza, Simone, Elias, Jaqueline, Nice, Andréa, Maria Aparecida (Nininha), and so many other very dear people, for all their co-operation and support.

The Earth, a welcoming mother, who sustains the lives of all those we love.

EPIGRAPH

"All things are impermanent

This is the law of births and extinctions

When both, life and death perish.

Nirvana will be the ultimate happiness"

Shinto Buddhist Teaching

SUMMARY

SUMMARY

The Atlantic Rainforest (MA) is one of the most threatened natural environments in Brazil, and its original area has been greatly reduced by anthropogenic actions. Agricultural activities are among the factors impacting this biome, and the region of Apiaí - SP, which produces table tomatoes (Solanum lycopersicum L.), a crop characterised by its importance for food production and responsible for significant socio-environmental impacts due to the use of agrochemicals, deforestation and poor working conditions, is part of the Atlantic Forest. Due to infestation by insect pests and the occurrence of agricultural diseases, this crop is considered one of the most demanding in terms of cultivation, requiring constant spraying and the general use of agrochemicals. This work seeks to expand knowledge about the functions that forest biodiversity can have as a tool to help agricultural management of phytophagous insects and diseases in table tomato crops. Between 2008 and 2010, five experimental tomato cultivation modules were set up in the municipality of Apiai - SP, using agroecological management in the model of High Productivity Islands (IAPs), in places where forest biodiversity is present in the surroundings. In each cultivation module, the following were analysed: forest biodiversity in the surroundings up to a distance of 300 m in the N, S, L, O directions; the occurrence of insect pests and agricultural diseases; and economic and socio-environmental viability. The results were compared to the same data collected from five other conventional crops in the same ecological region, which cultivated similar varieties over the same period in environments with less native forest biodiversity in their surroundings. The experimental agroecological crop modules showed an average area of Atlantic Forest cover of around 43.82%, with an average occurrence of 46 native tree species whose Shannon diversity index was 3.87. The comparative conventional crops had an average area of forest cover around 31.34 %, with an average occurrence of 26 native tree species and a diversity index of 3.44 for this category of species, which was significantly lower than that of the agroecological modules. In samples of 20 tomato plants per crop in the experimental areas, no viruses were found; a single species causing a fungal disease and a total of 4 species of insect pests were identified. The conventional producers compared the occurrence of 11 species of fungal and bacterial diseases, with an average of 1.4 plants/crop attacked by viruses and 7 species of phytophagous insects. The data showed that there is a relationship between forest diversity and the number of insect pest species in the crops. In an analysis of socio-environmental viability between the experimental and comparative crops, a higher rate of risk, investment, productivity, production volume and financial losses were identified, as well as greater environmental and social impacts among the comparative conventional tomato crops. The experimental models with forest biodiversity in their surroundings proved to be socially and environmentally viable for practical use in family farming in the region of Apiaí - SP.

Keywords: Forest biodiversity; High Productivity Islands; Agroecology;
Solanum lycopersicum; Plant health; Socioeconomics; Landscape ecology

CHAPTER 1

INTRODUCTION

The tomato (Solanum lycopersicum L.) is a vegetable of great economic importance in Brazil, mainly due to the volume of production and job creation, and is recognised as being of paramount importance in the diet of the Brazilian population (EMBRAPA, 1992). According to research by the Brazilian Seed and Seedling Trade Association (2010), tomatoes account for around 16% of the GDP generated by vegetable production in Brazil.

According to the Brazilian Institute of Geography and Statistics (IBGE, 2009), the area planted with tomatoes in the country in 2008 was 61,000 hectares, with production of more than 3 million tonnes, making it the most significant species, accounting for 20% of all vegetables produced. In the same year, tomato production in Brazil was responsible for generating 610,000 jobs (IBGE, 2009).

One of the characteristics of tomato production in Brazil is the migration of the crop in search of new areas, mainly due to the occurrence of phytosanitary problems that make production in a given location unfeasible. It is common to find areas that have been abandoned for long periods, more than 20 years, in order to "rest" and eliminate soil pathogens, mainly bacteria. Once the tomato-growing areas have been abandoned, they are occupied by pastures and annual crops as a way of "using up" the fertilisers and pesticides still present in the agro-ecosystem, without any forest restoration activities being carried out.

The tomato crop has a short cycle, is intensively mechanised and sprayed, and is difficult to manage due to the large number of insect pests and fungal, bacterial and viral diseases that occur during its cycle. As a result of the fact that it is extremely demanding in terms of climatic, environmental and soil conditions, tomato cultivation in the Apiaí-SP region is primarily carried out in recently deforested areas, with the aim of taking advantage mainly of the structuring of the environment and soils, their drainage and disease-suppressive capacity. Under these conditions, the crop has a very low incidence of disease, even when the pathogen is introduced, the plant is susceptible and the climatic conditions are favourable to its occurrence (RAGASSI, 2009)

The municipality of Apiai - SP, known as the gateway to the Atlantic Rainforest, is entirely within the Atlantic Biome and its forests are classified as high altitude semi-deciduous ombrophilous, with the presence of Araucaria (Araucaria angustifollia) (VELOSO, 1992). This municipality is home to one of the state's main table tomato production centres, and the base of this production chain is mainly made up of family farmers who manage this crop as their main source of income.

Due to the management strategy adopted by local farmers, the Apiaí region has suffered intense deforestation pressure, with major forest fragmentation in places where production centres are concentrated.

The current situation is one of increasing loss of the forest component, which implies loss of local biodiversity. At the same time, there has been an increase in the occurrence of pests and diseases in tomato growing, with an increase in costs, labour and scale of production. According to Gravena (2003), the occurrence and increase in the number of pests and diseases in crops can be related to the loss of native biodiversity and, consequently, of niches for natural predators of these harmful agents. This scenario generates a circle of degradation that is usually broken by the extinction of local productive resources, with the end result being the creation of degraded environments and impoverished communities.

In order to reverse this real and recurring situation in different agricultural systems, sustainable food production strategies or agroecological production models have been sought in recent decades.

Agroecological cultivation strategies advocate that the environment in which the crops are grown be "balanced" in such a way as to maintain characteristics that are suppressive to herbivores and pathogens that, whether or not they are present in the environment, will not cause significant infestations or diseases in agricultural crops in situations where there is a balance in the food chain. Some agroecological models aim to achieve a balance in the biological chains between phytophagous insects and their natural predators, through consortium techniques between the crops of interest and companion plant species that help disperse predators and promote other services of interest for the proper development of the desired crop. However, due to the aforementioned characteristics of the table tomato crop and its management, the need for internal intercropping with other plant species is a nuisance for farmers.

In the state of Acre, cultivation models have been developed for rubber trees (Hevea brasiliensis) in small cultivation modules within forest environments (KAGEYAMA et al., 2002). These models successfully advocate that small populations of this species, which in large plantations is susceptible to and decimated by the fungus Microcyclus ulei, which causes leaf blight, can be managed through the isolation provided by the forest structure as protection against pathogens.

In order to offer a more sustainable panorama for tomato production and income for producers in the Apiaí-SP region, as well as the possibility of increasing and maintaining the forest component in this region, the experimental cultivation system for tomato production in High Productivity Islands (IAPs) was developed. This system consists of managing small populations of tomato plants, under agroecological principles, on an economically viable scale, in environments that are at least 300 metres away from other similar crops, which must be separated from each other by the forest component.

Based on the above hypothesis, this study hypothesised that the unsustainable scenario of forest fragmentation linked to the high use of pesticides could be changed by using cultivation techniques in IAPs. In this proposed new scenario, maintaining the ecological services provided by the forest component leads to lower rates of disease and insect pest infestation in the tomato crop, reducing the use of fungicides and insecticides. This increases the economic efficiency of the crop, which can lead to lower rates of deforestation and greater conservation of water resources, as well as increased profits and improved working and health conditions for rural producers.

The aim of this research was to assess the influence of forest biodiversity on the occurrence of insect pests and diseases in two different tomato management systems in the municipality of Apiaí-SP: experimental cultivation on High Productivity Islands (IAPs) and conventional cultivation by local producers.

CHAPTER 2

LITERATURE REVIEW

2.1 Concept of Biodiversity

In recent decades, and especially since the 1980s, humanity as a whole has become more aware of the need to maintain the environment and natural ecosystems.

Grelle et al. (2009), indicate that since the 1990s, more than 380 articles have been published on the subject of conservation biology in Brazil, most of them related to biodiversity conservation in the Atlantic Forest and Amazon environments.

Various definitions of the term "biodiversity" have been proposed over the decades. The use and expansion of knowledge of this concept began in the 1980s, pioneered by Lovejoy (1980). In this study, Lovejoy reviewed various environmental and global systems considering energy, human populations, the economy and global forest resources and the consequences of their exploitation, such as climate change and species extinction estimates. Themes were written about biological diversity, formally creating an expression to define the number of species present.

Norse; McManus (1980 apud JEFFRIES, 1997; HARPER; HAWKSWORTH, 1995), who collaborated with the White House Council on Environmental Quality, in chapter 11 of the Annual Report of the Council on Environmental Quality, also during the mandate of President J. Carter, examine global biodiversity and define it with related concepts such as Genetic Diversity and Ecological Diversity, placing Ecological Diversity on the same level as species richness, in reference to the number of species in a community of organisms (JEFFRIES, 1997).

In both publications (JEFFRIES, 1997; HARPER; HAWKSWORTH, 1995), biodiversity was discussed on a global scale, related to broader themes and not just the biological aspect. With the current and potential importance of biodiversity recognised as the activity of natural ecosystems, providing what are now called vital services or functions for the health of Planet Earth, these documents made it clear that biodiversity should not be seen solely as an object of study for biology (JEFFRIES, 1997).

Biodiversity concepts currently consider the presence of biotic species and the complex environmental and social relationships involved in maintaining these components. These relationships, mainly due to the influence of anthropogenic activities, are currently identified as one of the main causes of biodiversity loss (EHRLICH, 1997).

According to Wilson (1997), tropical forests are the main areas of concentration of biodiversity, with only 7% of the planet's total coverage and containing more than half of the world's biota.

2.2 Atlantic Forest Biodiversity

Forest biodiversity is often cited as a structure or form of complex plant system that acts and has a positive and regulatory effect on various environmental functions that are beneficial to agricultural and forestry practices (ALTIERI; SILVA; NICHOLLS, 2003). On the other hand, there are examples that place forest biodiversity and its maintenance as a risk factor and an obstacle to agricultural development and production, pointing to the fact that forest biodiversity, its characteristics and functions, should be better studied in terms of their contribution to agricultural practices (ALTIERI; SILVA; NICHOLLS, 2003).

FAO figures from 2005 indicate that approximately 9.38 million hectares of native forests are lost every year worldwide, and that between 1990 and 2005, 64 million hectares of forests were deforested (FAO, 2007) throughout South America, with Brazil leading the way.

It is estimated that the forests of Latin America contain the greatest arboreal plant biodiversity on the planet, covering more than 1000 species, and also the greatest number of tree species considered threatened or vulnerable to extinction (FAO, 2007). According to Dean (1996), native forest structures are co-responsible for supporting and providing subsidies for the urban and agricultural development of the entire Brazilian nation (and others), through the use and exploitation of their resources over the five hundred years of Brazil's colonisation history.

One of the biomes targeted by deforestation in Brazil is the Atlantic Forest, which is also one of the most threatened on the planet, its area having been drastically reduced to 7% of the original in just a few centuries of occupation of this territory (SOS MATA ATLÂNTICA, 2010). Much of Brazil's social occupation is concentrated and superimposed on the original area of the Atlantic biome, whose original coverage was approximately 15% of the entire national territory and which, without its systemic characteristics maintained, represents a major loss of biodiversity for the entire planet.

This biome is home to approximately 260 species of mammals, 1020 species of birds, 197 reptiles, 340 amphibians and 350 fish, in addition to countless species of insects and other invertebrates, as well as hundreds of endemic plant species (RBMA, 2010). Heyer et al. (1988) recorded the supposed extinction of anuran species (Amphibia: Anura) in the Atlantic Forest region located in the Paranapiacaba continuum due to human action, and showed that the speed of degradation of this environment by anthropogenic action is much greater than the capacity of researchers and scientists to record and catalogue the species and ecological relationships contained in these ecosystems.

In Brazil, the decline of the Atlantic forest and its exploitation are associated with urban development, predatory extractivism and monoculture agriculture since colonial times (SMITH et al., 2008).

Agriculture and livestock farming are often fronts of impact on native forests, leading to a reduction and fragmentation of the areas occupied by these biomes, causing the extinction of species and ecological niches (ODUM, 2007) and profound impacts on the function and maintenance of these natural structures.

In recent decades, important work has emerged that demonstrates the need to preserve all national biomes, often culminating in significant actions by governments and society, such as the approval of laws and the creation of Protected Areas (DF 750/93; LF 9.985/2000).

The creation of Protected Natural Areas (LF 9.985/2000) and the development of legal instruments to protect forests are some of the strategies used to protect the Atlantic Forest biome, which are considered major advances in its conservation.

Altieri (2000) estimates that there are more than 16 million small farming families throughout Latin America, and places these people as co-responsible for predatory practices on native biodiversity, but points to the significant contribution that this group can make to environmental conservation if they adopt production and agricultural management techniques that are integrated and synergistic with local ecosystems.

In O mito moderno da Natureza Intocada (The Modern Myth of Untouched Nature), Diegues (1996a) argues that the behaviour of traditional communities and groups that live in and directly need the native resources of forests such as the Atlantic can often be an agent for protecting these ecosystems.

In another article, Diegues (1996b) argues that the conservation of Protected Natural Areas and Native Ecosystems must be allied to a socio-environmental pact, where natural environments are preserved and provide services and subsidies for the maintenance of human communities, which in turn can be active partners in the protection of these areas, combining agricultural productivity and extractive management with environmental conservation.

2.3 The regions of Apiaí and the Paranapiacaba-SP continuum

The history of Apiaí dates back to the early colonial period (18th century), when the region was exploited for its gold reserves and as a crossing point for troops of horsemen coming from the south of the country.

The municipality is currently known as the Gateway to the Atlantic Forest, as it is considered a region where the Atlantic Forest is highly preserved and in large fragments (Figure 1).

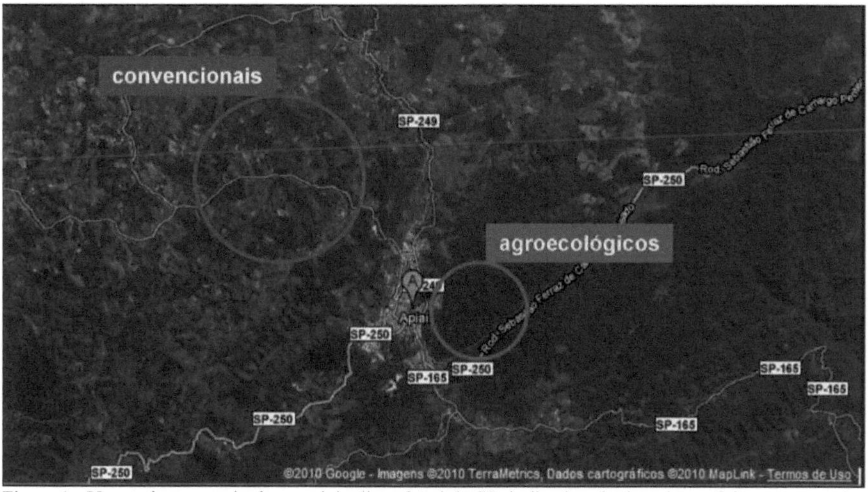

Figure 1 - Vegetation cover in the municipality of Apiaí - SP, indicating the locations of the crops monitored

By analysing the vegetation cover maps of the municipality, we can see that the largest forest remnants are in the east and southwest of the municipality. These regions are home to the Alto Ribeira Tourist State Park - PETAR, a Full Protection Conservation Unit considered to be one of the most biodiverse areas in the Atlantic Forest, with a rich natural heritage (SMITH et al., 2008) and an area of over 35,700 hectares (SÃO PAULO, 2010).

Another important remnant in the region is an area attached to PETAR, formerly known as Fazenda Vitória, which covers approximately 7,700 hectares and is currently the site of the "Edson Luis David de Macedo" Rural Settlement and Sustainable Development Project (PDS).

Next to these areas in the municipality of Apiaí, we can point to the presence of the "Morro do Ouro" Municipal Natural Park, created in 2003, with an area of over 540 hectares (SHIMADA, 2008). This park is considered an important remnant for maintaining the quality of the waters and springs that supply the municipality and guarantees one of the largest peri-urban green areas in the state of São Paulo.

In this way, these areas add up to more than 44,000 hectares of continuous forest, which characterises a mosaic of protected areas of great importance for the conservation of the Atlantic Rainforest, as they contain different formations of this biome, such as the High Altitude Dense Ombrophilous Forest with the presence of Araucarias and the Dense Ombrophilous Forest on Karst (VELOSO, 1992).

This mosaic of protected areas, which surrounds the municipality of Apiaí on the north-east, east and south-east sides, is added to the Paranapiacaba continuum, an area designated by the combination of the Alto Ribeira, Intervales and Carlos Botelho Tourist State Parks and the Xitué Ecological Reserve. Together they add up to more than 150,000 hectares of protected areas, which consolidates, along with all the other areas including the Jacupiranga Mosaic and the sustainable use conservation units, such as the Quilombos do Médio Ribeira and the Serra do Mar Environmental Protection Area, the largest remnant of Atlantic Forest in Brazil.

The municipality of Apiaí has historically had a strong family farming component, with an estimated 10,514 of its more than 25,700 inhabitants (IBGE, 2009) living in rural areas and basically using annual crops and livestock as a source of income and subsistence (IBGE, 2001) (Annex A).

2.4 The panorama of the deforestation front in the Apiaí-SP region

On its northern and western sides, the municipality of Apiaí has a large deforestation front, with the presence of large concentrations of reforested massifs and the use of several small areas for intensive agricultural crops and pastures. This region is located under the domain of the Atlantic Forest cover (DF 750/93 and CONAMA RESOLUTION 10/1993), considered one of the richest forest formations in diversity, with a panorama of occupation and deforestation similar to that of other regions and biomes in the country (RBMA, 2010). In this way, we can cite local agricultural practices in the Apiaí - SP region as a factor in the impact and reduction of biodiversity in environments originally occupied by forest formations.

2.5 Tomato growing and its presence in the Apiaí region

The tomato (Solanum lycopersicum L.) is one of the most important vegetables consumed and cultivated all over the world and is part of the typical cuisine of various ethnic groups on the planet, occupying first place in volume of vegetable production in Brazil (SILVA et al., 2003; SILVA; LOPES; MAGALHÃES, 2010). Due to the diversity of shapes, sizes, colours and flavours, tomatoes have a wide range of uses (CONSUEGRA et al.; 2000)

As mentioned by Silva et al. (2003), it is the most important vegetable crop in Brazil and is grown in all regions of the country. Since the 1950s, it has gained importance in the south-east and in the state of São Paulo. This crop is responsible for the direct employment of more than 200,000 people in the country (TAVARES, 2003).

According to Barbieri; Stumpf (2008), tomato cultivation is distinguished by a number of favourable characteristics, including nutritional properties such as the presence of carotenoids (SILVA et al., 2003), vitamin C, potassium, as well as proven action in the prevention of prostate cancer, vascular diseases and eye damage caused by UV rays (FAGUNDES et al., 2005).

According to Agarwal; Rao (2000 apud BARBIERI; STUMPF, 2008), due to the presence of lycopene, a carotenoid with antioxidant properties.

Tomatoes grown for fresh or table consumption are considered to be of indeterminate growth (RICK, 1995) and need to be trained and harvested by hand and staggered, while tomatoes for industrial processing are of determinate and creeping habit, with the fruit used in agro-industry (BARBIERI; STUMPF, 2008).

According to Barbieri and Stumpf (2008), one of the difficulties in producing table tomatoes is their perishability, which forces them to be marketed in a short space of time. They are one of the crops with one of the highest rates of post-harvest losses, mainly due to inadequate handling during harvesting and transport.

Currently, table tomato producers have opted for so-called long-life varieties which, by keeping the pericarp of the fruit firm for longer, significantly increase its post-harvest preservation, allowing it to be transported between more distant regions (PIERRO, 2002).

The tomato plant originated in the territory between Ecuador, Chile, the Pacific Ocean and the Andes Mountains, in places with moderate rainfall and average temperatures of around 15° to 19°C (SILVA et al., 2003). The cultivated species, Solanum lycoperiscum, originated from the Andean wild species Solanum cesariforme, which produces cherry-type fruit (RICK, 1995).

The tomato was domesticated in the region of Mexico, outside its place of origin, where it was taken in pre-Columbian times by trade routes between Andean and Mexican peoples. In the region where it was domesticated, it was grown in consortium with maize, beans and pumpkin plants (CONSUEGRA et al., 2000).

With the arrival of the Spanish colonisers in Mexico, they found tomatoes being widely consumed by the local population and took them to Europe, where they spread to various countries such as England, Italy and Spain. However, for a long time, the tomato was considered an ornamental plant and its fruit was only widely consumed around the 19th century (RICK, 1976).

In Brazil, tomatoes were introduced by European immigrants at the end of the 19th century (ALVARENGA, 2004).

In the Sorocaba region, in the municipalities of Guapiara and Apiaí - SP, tomato cultivation began in the 1950s (MINAMI, 1980) and to this day it is the most important table tomato producing region in the state of São Paulo.

Due to the fact that the municipality of Apiaí has a winter period with frosts, tomato growing in the region takes place between the months of August and March. Due to its subtropical climate with temperate characteristics, it has the best conditions for tomato growing (FILGUEIRA, 1972). Although these geographical and environmental conditions in the region are, for the most part, favourable to tomato growing, the average annual rainfall is above the ideal recommendations for this crop, which, for example, makes commercial cultivation possible only with the adoption of control methods for late blight (Phytophthora infestans) (MINAMI, 1980).

In this municipality, tomato cultivation is one of the main production strategies of family farming, as it suits the profile and density of occupation of this territory. According to data from the Apiaí Rural Union, more than one million tomato plants were grown during the 2009 harvest. Most of the production is provided by family farmers who, in agreement with entrepreneurs or local traders, have their crops financed in conjunction with the obligation to supply their harvest to the partner at fluctuating prices dependent on the CEASA - SP commercialisation peaks.

In the conventional tomato cultivation system, there has been a significant increase in the occurrence of pests and diseases in these crops (PRIMAVESI, 1988), which has forced many of the local farmers who adopt this cultivation system to invest large sums of capital in pesticides and human resources in order to maintain their productivity.

Tomatoes are traditionally a crop that is highly vulnerable to the occurrence of pests and diseases, and their list of pathogens includes more than 30 possible types of disease (KUROZAWA; PAVAN, 2005) and 15 potential insect pests (GALLO et al., 2002).

Currently, a conventional grower in the municipality of Apiaí - SP subjects his crop to an average of approximately 36 chemical sprays per cycle (TOMAS et al., 2009), which makes tomato growing one of the most expensive and contaminating agricultural systems, both for the environment and for the people involved in its production and consumption.

On a recurring basis, we have seen local producers harvesting large tomato crops and selling them to urban markets, but facing financial losses due to high production costs and low prices, as a local agricultural technician said in an interview: "For decades, producers have not been able to make a good profit from tomato production."

2.6 The creation and development of sustainable alternatives for the conservation of the Atlantic Forest and the socio-economy of regional family farming

In recent decades, since the green revolution, we have seen an increase in studies and the development of agricultural techniques that are passed on to farmers and which preach the use of synthetic and chemical inputs to maintain soil fertility levels and control and prevent the action of insect pests and diseases. In these technological packages, we can see the proposal to replace forest environments in favour of cultivated areas, which, after initial management, remain without native forest biodiversity and will be destined for a cycle of crop rotations primarily with pastures and annual and perennial crops. In recent decades, this technological model has led to technicians and farmers "distancing" themselves from the knowledge and use of forest environments, and in this period they have been partially and progressively ceasing to use these forest ecosystems for the benefit of agricultural production (ALTIERI, 2002).

In this context, with rural producers facing rising production costs due to the increase in insect pests and diseases in their crops on the one hand, and drastic falls in forest cover and native biodiversity in the region on the other, the question arises as to whether the indicators of these two situations could have a direct relationship of interaction.

In recent years, this situation has become evident in different agricultural production centres, which has led to the emergence of various initiatives to develop more sustainable cultivation techniques .

At the same time, there has been a growing concern throughout society about the impact of agriculture on the environment and the contamination of the food chain with pesticides, resulting in the presence of market segments looking for differentiated products, free from pesticides and practised in socio-environmentally sustainable crops (BETTIOL; GHINI, 2003).

In this sense, we can mention some successful strategies for maintaining agricultural productivity at low levels of inputs and production costs, using biological and environmental services, such as some models of agroforestry systems, like the RECA project (SILVA, 2006) among others (VIVAN, 2003) and the strategies of High Productivity Islands, which advocate small cultivation centres isolated from each other by the forest component, tested and in production throughout Brazil and the world.

Lutzemberger (2002) mentions that changes from agrochemical management to agroecological nutritional management of orange orchards led to the suppression of the pathogenic characteristics of the bacterium Xanthomonas citrii, which causes citrus canker, which remained in the environment but did not manifest itself as an infectious agent in the crop.

Kageyama et al. (2002) demonstrated the possibility of obtaining productivity and health in rubber trees grown in a High Productivity Island system in the state of Acre, with healthy plants, without the need to control pests and diseases, in areas that are unviable for large-scale cultivation due to the occurrence of the causal agent of leaf blight, the main cause of commercial crop failure in the Amazon (MATTOS et al., 2003). In these agroforestry systems and IAPs, the strategies point towards smaller-scale cultivation than conventional ones, basically advocating agroecological management and consortium with various tree species, based on replacing synthetic inputs with biological and natural promoters of the regulation and maintenance of agricultural productivity.

The work of family farmers involves a variety of functions, such as food production and environmental and cultural services, proving the multifunctionality of their socio-environmental role and thus, as Diegues (1996b) points out, they can represent a potential agent for helping to develop mechanisms to prevent or minimise the loss of native and Atlantic biodiversity.

With the integration of social dynamics, motivating innovations and the development of agroecological production models, regardless of the credibility of these techniques in the academic world, historically marginalised groups of small rural producers have been generating their own benefits aimed at conserving the ecosystems in which they live and produce.

Considering the hypothesis that forest biodiversity can be an important and effective tool for promoting plant health in agricultural crops, this study aims to test the efficiency of the experimental agroecological model in High Productivity Islands for tomato cultivation, analysing the use of the forest component as a tool for regulating the occurrence of insect pests and diseases. To this end, the occurrence and severity of diseases, as well as the population of insect pests, were analysed in tomato crops in the municipality of Apiaí-SP, in experimental agroecological management systems in IAPs and comparative conventional management, in environments characterised by high and low forest diversity in their surroundings, respectively. During the cultivation of the crops, financial and socio-environmental information was collected to generate indicators of the socio-economic viability of the experimental models.

CHAPTER 3

MATERIAL AND METHODS

3.1 Experimental Proposal

Seeking to fulfil the function of research aimed at building social thinking (FREIRE, 2004), the project was conducted in real field situations, with the participation and support of local and family farmers in the municipality of Apiaí - SP, where experimental tomato cultivation modules were developed in an agroecological management system on 5 local properties with a predominance of Atlantic forest in their surroundings. At the same time as the experimental crops were being grown, the crop was also cultivated in a conventional management system on 5 properties in the same region with similar socio-environmental conditions, which served as a comparative benchmark for the indicators analysed.

All 10 crops, 5 agroecological and 5 conventional, had their surroundings analysed for plant biodiversity and forest cover. In each crop, the occurrence of insect pests and diseases was determined in order to compare the agroecological and conventional management modules, as well as to verify the relationship between plant biodiversity in the surroundings and the occurrence of pests and diseases.

At the same time as the crops were grown, social and economic information was collected to support the generation of indicators on the socio-environmental and economic viability of the two management systems, relating this to the practical applicability of using forest biodiversity as an auxiliary tool for agricultural production in table tomato crops.

3.2 Description of the proposed experimental models

The sites for the experimental agroecological cultivation modules, as well as the comparative conventional cultivation modules, were chosen on the basis of similar soil, climatic and social conditions, with the inputs used during cultivation, the form of management and the forest biodiversity around the crops being the distinguishing features.

Over the course of two years, a total of 10 tomato producers in the municipality of Apiaí - SP took part in this project.

3.2.1 Experimental agro-ecological cultivation modules with high forest biodiversity in the surroundings

Technical support for the producers responsible for running the experimental modules was provided by students and agronomists linked to the Agrarian Reform Agroecological Development Project (PPDARAF)/USP.

The technical recommendations given to these farmers are based on the principles of Agroecology (GLIESSMAN, 2001) and Sustainable Agriculture (PRIMAVESI, 1992), using primarily techniques aimed at the interaction of crops with Forest Biodiversity and the Ecological Management of Pests and Diseases (VIVAN, 1998; PRIMAVESI, 1988), as a way of enabling the natural regulation of insect pests and diseases in crops. The inputs used, when necessary, in the agroecological crops were certified and were made available in equal quantities to the five producers taking part in the project.

The Sustainable Development Project (PDS) Rural Settlement - "Luiz David de Macedo", located in the Caximba neighbourhood in the municipality of Apiaí-SP is an area of approximately 7,700 hectares, with at least 5,500 ha of this property under Atlantic forest cover, home to approximately 83 farming families (SANTOS et al., 2008).

The producers were selected in collective assemblies of the Settlement's Rural Producers' Association, using participatory and community methods with a conservationist and organisational focus (OOSTRUM, 1996; OOSTRUM; STIVEN, 1997).

The experiment lasted two agricultural cycles during 2008 and 2010, with two farmers from this settlement taking part in the first year and three farmers in the second year (Table 1).

The settled farmers participating in the experimental project were referred to as agroecological farmers when they were mentioned together, and their cultivation areas were classified as agroecological experimental modules.

The principle of cultivation in High Productivity Islands (KAGEYAMA et al., 2002) was used, with forest biodiversity as the main component surrounding the tomato crops. As mentioned by Altieri et al. (1981), the influence of biodiversity in the surroundings has a range of 40 metres from the edge to the interior of the crop, and this was the criterion used for sizing the experimental modules, which had dimensions of approximately 25 x 25 m and a maximum total of 1000 tomato plants of the Débora variety, planted at a spacing of 0.7 x 1.0 m (Table 1). Sowing was done in trays in nurseries and the seedlings were transplanted 45 days after sowing.

The average distance between the experimental crops was over 300 metres, with at least 3 km of distance from conventional crops that could interfere or cause contamination from pesticide drift and contamination of watercourses.

Table 1 - Location (latitude, longitude), altitude, cultivated area, number of plants and sowing time of tomato variety Débora in 5 experimental agroecological cultivation modules in the municipality of Apiaí-SP, Caximba neighbourhood.

	Experimental modules				
	A1	A2	A3	A4	A5
Property	A1-3	A2	A1-3	A4	A5
Producer	Zacchaeus	Joaquim	Zacchaeus	Toninho	Zé Pinheiro
Latitude	24°27'38.41"S	24°28'36.48"S	24°27'38.41"S	24°27'41.52"S	24°28'09.68"S
Longitude	48°46'47.23"O	48°46'53.02"O	48°46'47.23"O	48°47'55.64"O	48°47'34.40"O
Altitude	900 m	880 m	900 m	930 m	930 m
Area cultivated	600 m^2	523 m^2	656 m^2	698 m^2	698 m^2
number of plants	860	750	940	1000	1000

Sowing time	Oct 2008	Oct 2008	Sep 2009	Sep 2009	Sep 2009

3.2.2 Comparative modules of conventional cultivation with low forest biodiversity in the surroundings

The rural neighbourhood of Encapoeirado is home to several family farmers who work professionally growing table tomatoes.

Due to the history of occupation in the area and intensive agricultural practices, this neighbourhood has a fragmented forest landscape, highly impacted by the pressure of agricultural activities.

Two producers took part in the project in 2008, and three in 2009 and 2010, who were referred to as conventional farmers (Table 2).

Table 2 - Location (latitude, longitude), altitude, cultivated area, number of plants and sowing time of tomato variety Débora Max, in 5 conventional modules in the region of Apiaí-SP, Encapoeirado neighbourhood.

	Experimental modules				
	C1	C2	C3	C4	C5
Property	C1	C2	C3	C4	C5
Producer	Sérgio Barbosa	Farm Reunited	Ivo	Isaltino	Adam
Latitude	24°30'00.00"S	24°27'47.56"S	24°27'43.89"S	24°26'35.80"S	24°28'36.14"S
Longitude	48°48'01.00"O	49°00'52.62"O	48°55'08.80"O	48°54'47.73"O	48°55'3.95"O
Altitude	906 m	820 m	940 m	925 m	890 m
Area cultivated	34.998 m^2	13.298 m^2	6.998 m^2	8.398 m^2	24.498 m^2
number of plants	50000	19000	10000	12000	35000
Planting time	Oct 2008	Sep 2008	Sep 2009	Sep 2009	Sep 2009

The seeds were sown in trays in nurseries and the seedlings were transplanted 45 days after sowing. The spacing used was 0.7 x 1.0 m

The agricultural management adopted by producers in places with low forest biodiversity in the surrounding area involves the use of agricultural inputs to aid production, such as synthetic fertilisers, herbicides, insecticides, fungicides and others.

Producers who work under conventional tomato production are monitored by local technicians, and their crops are observed and analysed in comparison with the experimental modules proposed.

The data on the occurrence of insects and diseases was obtained from crops with effective control of these indicators and aimed at optimising production.

3.3 Characterisation of forest cover and plant biodiversity around crops

The growing areas of all the producers involved, both agroecological and conventional, had the presence of the plant component in their surroundings characterised according to the methodology developed and suggested by the technical and research group at LARGEA/ESALQ/USP.

The analysis of the biodiversity of the forest component around the tomato crops was carried out by sampling linear transects, each 300 metres long, in the north, south, east and west directions, established from the centre of the agroecological crops and, in the case of the conventional crops, because they are larger, from the extreme edges in each direction.

In each transect, around 6 sampling points were marked, with a distance of 50 metres between each point, in a total area of approximately 300,000 $metres^2$ in the surrounding area. The sample plots were 2x2 metres in the case of sites with herbaceous and shrubby vegetation, and 5x5 metres in sites with arboreal vegetation, in order to ensure greater representativeness. The points were established with the aid of tape measures, photographed and georeferenced (Garmin Etrex GPS devices) for the subsequent preparation of composite maps using satellite images.

All the plant species were collected from each plot and kept in exicates for later identification in the Botany laboratory at ESALQ/USP, which allowed the panorama and sampling profile of the forest cover situation around the crops to be structured.

The frequency of plant species around each crop (property) was calculated as the total number of individuals in relation to the total number of species found.

The absolute results obtained in the plots of the experimental and comparative producers, relating to Forest Biodiversity, were converted into Shannon Diversity indices (SHANNON, 1949), according to the formula:

$$H' = -\sum_{i=1}^{S} pi \ln pi \; ;$$

Where H'= Shannon index; S= number of species; pi= relative abundance of each species (occurrence of species/total species).

Indices were calculated for each category of species (native and exotic) in agroecological and conventional crops, considering all the species and individuals sampled on the 5 properties. The diversity indices were compared using the t-test as proposed by Hutcheson (1970 apud ZAR, 2010).

Maps were drawn up locating the sampling points and indicating the rates of native forest cover in the perimeter sampled around the crops of each of the 10 producers, as well as comparative tables and graphs indicating the occurrence and frequency of plant species.

The Google Pro programme was used to estimate the area of forest cover around 300,000 m^2 of each crop.

3.4 Occurrence of insect pests and diseases in agroecological and conventional tomato crops

Between 2008 and 2010, we monitored the crops of 10 tomato growers whose management differed between the conventional model with the use of agrochemicals and the agroecological model without the use of synthetic inputs.

The crops were monitored from the time the tomato seedlings were transplanted into the field until they were harvested. Three monthly assessments were made per crop cycle, the first 20 days after planting the seedlings, the second at the start of fruiting (45 days after planting) and the third during harvest, approximately 70 days after planting.

The technical monitoring took place on a weekly, fortnightly or monthly basis using different methods adapted to each research indicator.

Data on these occurrences was obtained using the sample collection method, with 20 plants in the productive phase per crop being randomly selected for each evaluation and analysed throughout their structure.

The phytophagous insects were counted in terms of the number of individuals per plant sampled, using magnifying glasses and notes on field cards. The values presented per producer were obtained from the arithmetic mean of the sum of the number of insects identified per plant sampled divided by the total number of plants sampled.

Because tomato fruits infested with the small fruit borer, Neoleucinodes elegantalis, are discarded immediately, regardless of the number of larvae infesting the fruit, the data on the occurrence of this insect is based on the number of fruits attacked x number of insects per fruit, the latter coming from the count on one fruit from the respective plant.

Diseases caused by bacteria and fungi were identified and quantified in terms of their presence or absence in the crop (MALAVOLTA JR., 2004). The occurrence of viral diseases was quantified as the number of diseased plants in 20 plants sampled. In the case of late blight, caused by the fungus Phytophthora infestans, the percentage of affected individuals and the severity of the infestation (leaf area affected) were determined using the severity key proposed by James (1971).

Because the occurrence of various fungi and bacteria in tomato plants is associated with drainage conditions and soil structure, soil compaction was assessed in both the agroecological and conventional modules using a Hatô penetrometer, which measures the depth (up to 90 cm), in centimetres, of compacted layers with a resistance greater than 2.0 MegaPascal (RAGASSI, 2009). To do this, 5 points were randomly and homogeneously selected for each crop.

The occurrence of viruses in plants was determined by the presence or absence of symptoms, which is directly related to the occurrence of insect vectors (GRAVENA, 1984).

The results for each management system, agroecological and conventional, come from the arithmetic mean of the crops evaluated.

Shannon Diversity Indices (SHANNON, 1949) were also calculated, as described above, for the insect pests in each of the crops, considering a total sample of 100 tomato plants.

3.5 Analysing the relationship between forest biodiversity around tomato plantations and the occurrence of insect pests in the crop

To assess the relationship between tree and plant biodiversity and the occurrence of insect pest species, the correlation test with t-Student distribution (gl=n-2) was applied.

3.6 Socio-environmental and economic analysis

Comparative agroecological and conventional tomato crops had their socioeconomic indicators sampled in order to determine the environmental and social viability of the proposed models.

The conditions under which the modules were studied were real in the field, where all the social and environmental factors that can influence laboratory research were present, such as environmental inclement weather, price and market fluctuations, as well as the physical and human condition of those involved in the day-to-day work of cultivation.

The aim was to estimate the main costs involved in tomato production under the two management proposals, their total productivity, harvest utilisation, volume marketed or used, as well as the sales values and profits obtained during the harvests. This information was collected through interviews and day-to-day contact with the producers, and notes were made on field records. The permanent equipment involved in growing tomatoes, such as tractors, irrigation systems, agricultural implements and other vehicles, for example, were not included in the costs of the production activity, and their depreciation and acquisition were considered to be sources of an increase in total production costs.

Productivity and revenue calculations were based on Samuelson; Nordhaus (1993).

CHAPTER 4

RESULTS AND DISCUSSION

4.1 Characterisation of forest cover and plant biodiversity around tomato plantations in the Apiaí-SP region

a) Experimental agroecological modules

Due to the fact that the rural producers of the "Luiz David de Macedo" PDS have been settled since 2007, some of the areas occupied by the producers are in the process of regenerating, with a wide range of pioneering species of forest regeneration close to their cultivation areas.

Module A1 - 3

From a central point located in the tomato crop, 24 sampling points were established (Figure 2), 10 of which coincided with areas of native forest cover and 14 with areas without native forest cover. Of the total area (300,000m^2), 33.3% has native forest cover, which corresponds to 99,883 m^2. In the case of this property, there are several areas undergoing regeneration, the tomato plantation is located in the centre, and on the edges there are several structured forest fragments.

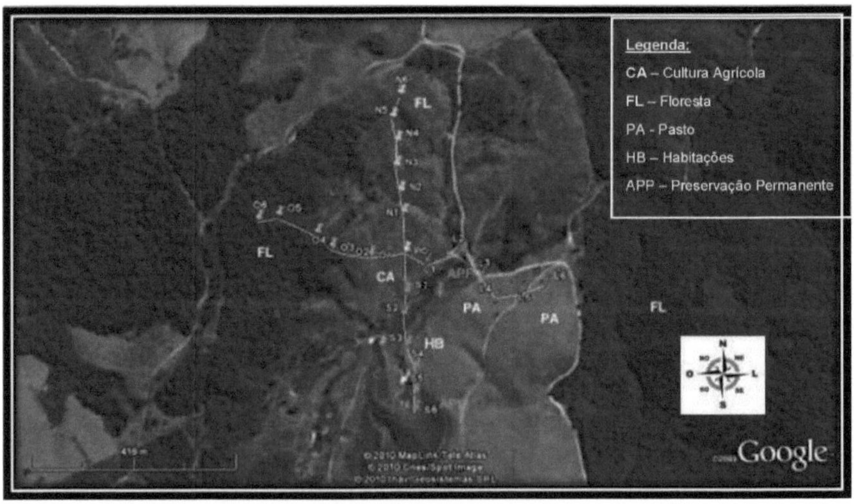

Figure 2- Sampling points for characterising forest biodiversity in cultivation module A1-3

Module A2

From a central point located in the tomato plantation, 24 sampling points were established (Figure 3), 16 of which were in areas covered by native forest and 8 of which were not. Most of this area (76.5%) is covered by native forest (229,468 m^2), and the tomato-growing area is completely surrounded by forest vegetation. The map shows that near the SP-250 motorway, looking south, the land is devoid of forest vegetation.

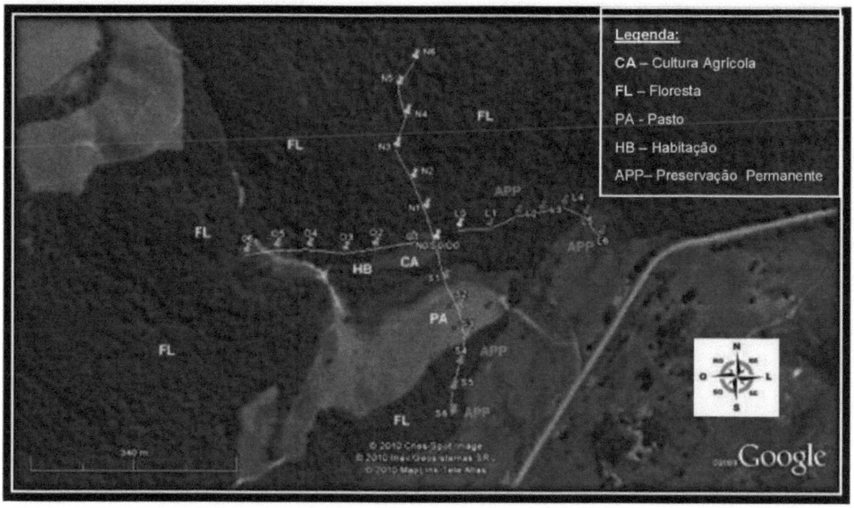

Figure 3 - Sampling points for characterising forest biodiversity in cultivation module A2

A4 module

From a central point located in the tomato crop, 24 sampling points were established (Figure 4), of which 7 and 17 coincided in areas covered and not covered, respectively, by native forest. On this property, there are several agricultural crops near the tomatoes, such as pole beans (Vigna unguiculata [L] Walp), sorghum (Sorghum bicolor [L.] Moench), yacon (Smallanthus sonchifolius [Poepp. & Endl.] H. Robinson); and also a well-structured area of Atlantic forest, rich in lianas (such as the manila liana, Aristholochia triangularis Cham) and with the presence of sugar fern (Cyathea sp) and important native trees, such as black cinnamon (Ocotea catharinensis Mez), caixeta (Tabebuia cassinoides [Lam.] DC.) and others. The area of native forest cover is 69,623 m^2, corresponding to 23.2% of the total sampled (300,000 m^2)

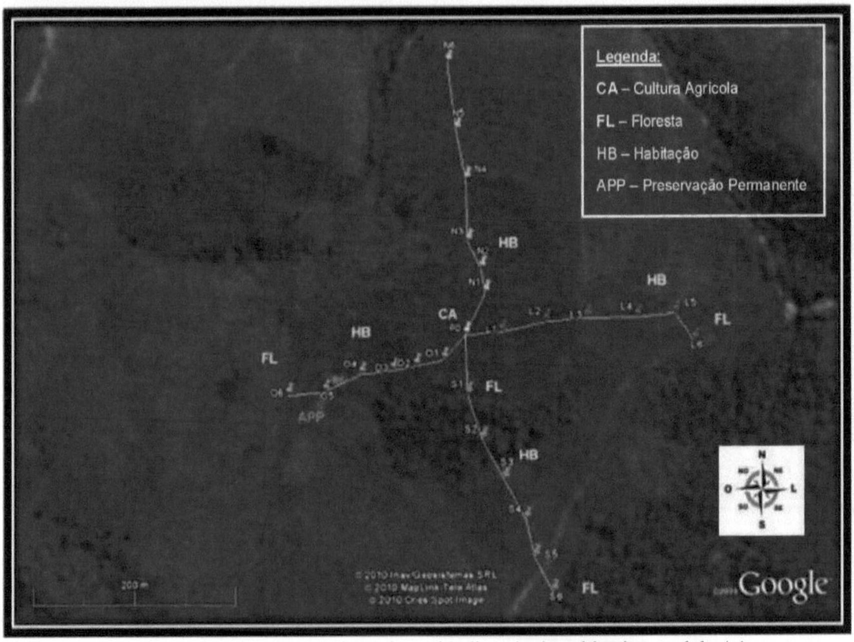

Figure 4 - Sampling points for characterising forest biodiversity in cultivation module A4

Module A5

From a central point located in the tomato crop, 24 points were established (Figure 5), of which 8 were in and 16 were not in native forest areas. In this module, a large part of the area used by the producer is occupied by brachiaria grass (Brachiaria decumbens Stapf), which is in the process of regeneration. With good plant diversity in the areas occupied by the forest component, in watercourses and small fragments, it can be seen that in addition to the area sampled in transects (300,000 $m^{2)}$, there are large continuous remnants of Atlantic Forest formation. The area of native forest cover was estimated at 158,627 m^2, 52.8 per cent of the total.

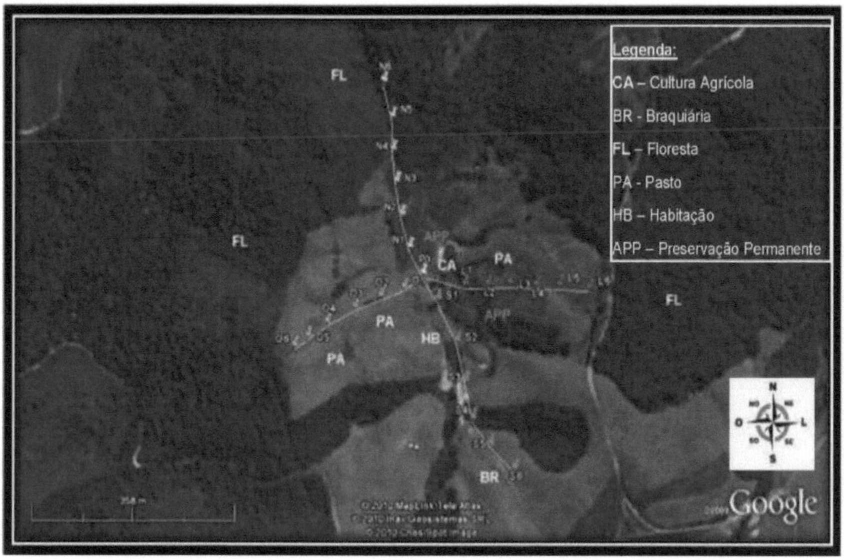

Figure 5 - Sampling points for characterising forest biodiversity in cultivation module A5

b) Comparative conventional modules

Module C1

A total of 26 points were sampled (Figure 6), of which 7 and 19 coincided in areas covered and not covered, respectively, by native forest. The area of native forest cover was estimated at 118,549 m^2, 39.5% of the total.

There is a forest fragment to the north-east of the cultivation site, very close to a road that borders the cultivation site. To the south, there are other tomato plantations less than 300 metres away. Between these two areas of tomato cultivation is an APP area, with springs and the formation of a small watercourse that could be used to supply sprayers, irrigation and be contaminated by the use of agrochemicals nearby.

Figure 6 - Sampling points for characterising forest biodiversity in cultivation module C1

Module C2

A total of 27 sampling points were established (Figure 7), of which 6 were in and 21 were not in areas of native forest. The area of native forest cover was estimated at 61,970m^2, 20.65% of the total. The area around the C2 module is characterised by a very small forest component, impacted by the colonisation of grass plants and restricted to permanent preservation areas around watercourses. One problem pointed out by conventional producers is the drift of agrochemical products from the cultivated areas into the forest fragments, which can lead to a reduction in the natural occurrence of plant species.

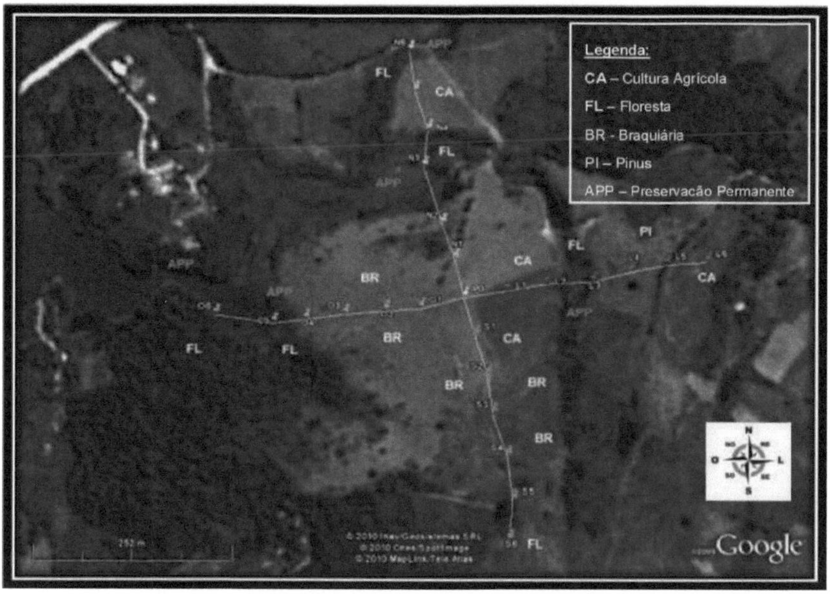

Figure 7 - Sampling points for characterising forest biodiversity in cultivation module C2

C3 module

Twenty-six sampling points were marked around this crop (Figure 8), of which 5 and 21 coincided in areas covered and not covered by native forest, respectively. The area of native forest cover was estimated at 41,096 m^2, 13.7% of the total, and is located close to watercourses. Most of the sampling points were in areas characterised by a low quantity of plant species in their composition, being predominantly occupied by spontaneous herbaceous species and annual crops such as maize and beans.

Most of this sampling area has a high profile of fragmentation, with forest remnants being dispersed and isolated, which hinders gene flow and the maintenance of reproductive processes between tree species that are contained both inside and on the edges (KAGEYAMA; GANDARA; SOUZA, 1998).

Figure 8 - Sampling points for characterising forest biodiversity in cultivation module C3

C4 module

Twenty-four sampling points were established (Figure 9), of which 4 and 20 coincided in areas covered and not covered, respectively, by native forest. The area of native forest cover was estimated at 66,600 m^2, 22.2 per cent of the total.

This area is almost entirely characterised by environments that have been altered by anthropogenic activity, with many crops of maize, beans and eucalyptus.

The forest remnants present in this area are quite altered, with selective felling of trees in their interior, signs of afforestation, the passage of domestic animals and human walking. Despite this panorama of impacts on the presence of the forest component, we can see the presence of adult individuals of different tree species up to 30 metres tall, which makes these fragments important from the point of view of providing seeds and maintaining local ecological processes.

Figure 9 - Sampling points for characterising forest biodiversity in the C4 cultivation module

C5 module

27 sampling points were established (Figure 10), of which 12 of which were in areas with forest cover and 15 in open areas.

Most of the forested areas are made up mainly of secondary forest with small diameter trees and the presence of liana and pteridophyte species. The structure of the southern transect is made up of a single forest fragment, diversified in tree species with the presence of

Araucaria, but with several internal paths, a dammed watercourse and selective cutting of some tree species. The other sampling points with forest cover had various impact indicators such as afforestation, selective cutting of species, walking and trails for domestic animals, but most were bordered by annual crops, pastures or stands of exotic trees (Eucaliptus spp. and others). The area of native forest cover was estimated at 182,033 m^2, 60.67% of the total sampled.

In the open areas there was a low diversity of species, most of which were herbaceous, with few shrub species and no tree plants.

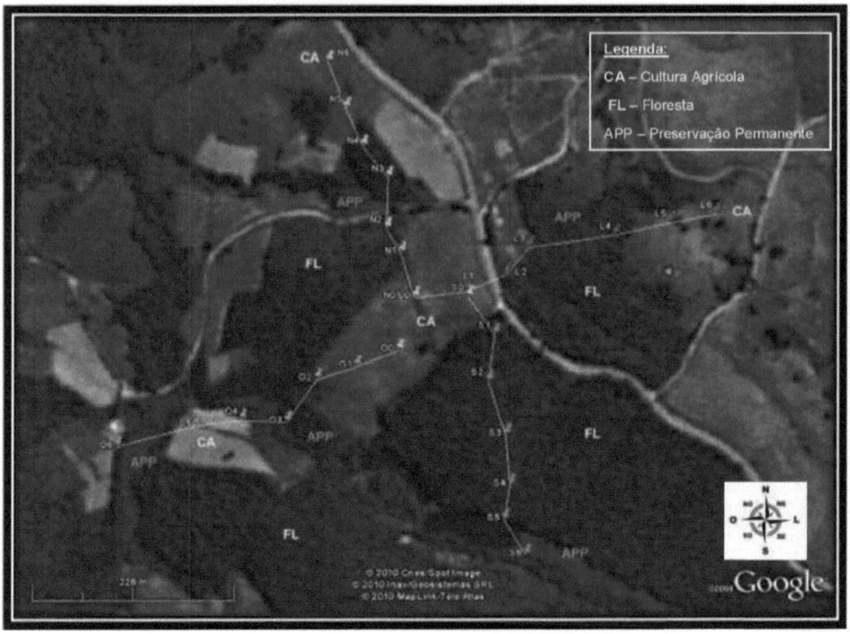

Figura 10 - Sampling points for characterising forest biodiversity in cultivation module C5

In the surroundings of the agroecological crops, 45.84% of the plots had forest cover, and in the conventional crops, 26.26%. The average area of forest cover around conventional tomato crops was 94,049 m^2, which represents 31.4% of the total (300,000 m^2) (Table 3). The higher rate of forest cover among the agroecological crops was confirmed by analysing satellite images (Google Pro) and between the sampling points.

Analyses of the images of the same crop fields show a higher rate of forest fragmentation and the presence of pastures and annual crops, which together with lower levels of native plant biodiversity, according to Panizzi; Parra (1991), help to create refuge environments for insect pest species.

Table 3 - Characterisation of forest biodiversity in the surroundings (300,000m^2) of experimental and comparative conventional agroecological tomato crops in the municipality of Apiaí-SP. Mean ± SD (N=5)

		Crops	
		Agroecological Experimental	Conventional Comparisons
Native forest cover	Area (m^2)	131.496 ± 28.426,78	94.049 ± 25.424,36
	%	43,82 ± 9,47	31,34 ± 8,47
Average area infested by grasses and other invasive plants (m^2)		15,5 ± 3,66	49,3 ±12,51

The surroundings of the experimental agroecological crops had a higher average number of tree species than the surroundings of the conventional crops, 46.75 and 20.60 respectively (Figure 11). However, the number of trees was practically the same for both crops (Figure 12), indicating that the forests between the agroecological producers are well established, with greater species richness, larger trees and higher dispersal rates between plants.

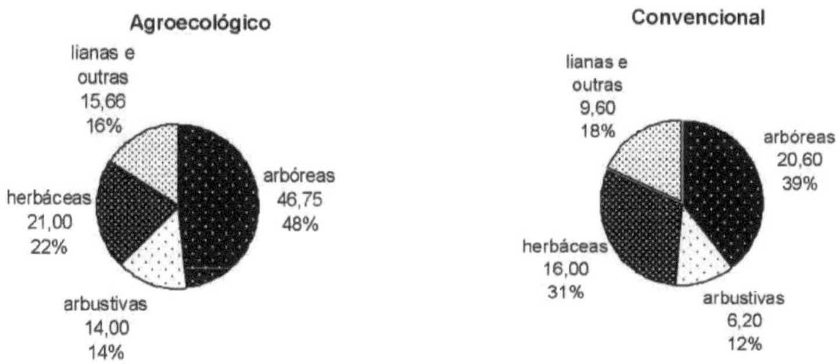

Figura 11 - Number of plant species sampled around experimental agroecological and conventional tomato crops (300,000 m^2) in the municipality of Apiaí-SP. Average values (n=5)

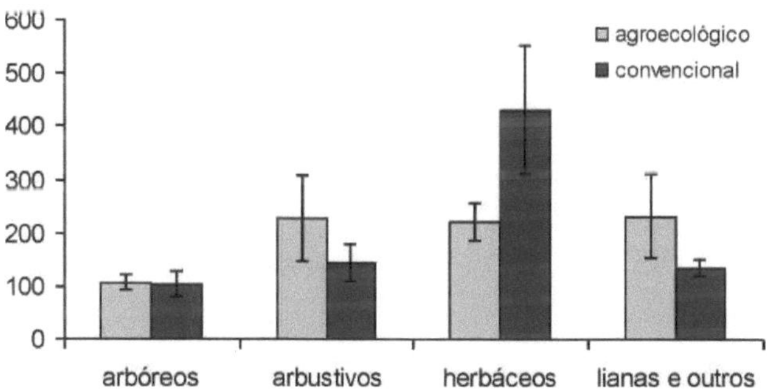

Figura 12 - Number of plant individuals sampled around experimental agroecological and conventional tomato crops (300,000 m^2) in the municipality of Apiaí-SP. Average values (N=5)

The regenerating areas on the agroecological properties proved to be more biodiverse in terms of shrub species, with a higher number of individuals than the comparative conventional areas, probably due to the fact that these areas are in an early stage of forest regeneration.

The surroundings of the agroecological crops had a greater diversity of native tree species than the conventional ones ($P<0.05$) (Table 4). There was no difference between the crops for the other categories. All the exotic tree and herbaceous species found are, in their entirety, of interest or other agricultural crops, such as medicinal and food plants, e.g. avocado (Persia americana), lemon (Citrus spp.), banana (Musa spp.), string bean (Vigna unguiculata L. Walp.) and others.

Table 4 - Shannon Diversity Indices (H') and number of species for categories of different vegetation strata in the surroundings (300,000 m^2) of tomato crops in experimental agroecological and conventional comparative management systems

Category of species	Agroecological		Conventional		t tabulated	t calculated	
Arboreal	H'	number of species	H'	number of species			
Native	3,8750	60	3,4435	34	1,900	2,6022	*
Exotic	1,0397	3	0,6816	2	2,262	0,5202	ns
Shrubs							
Native	2,3007	17	2,2143	9	1,960	1,2586	ns
Exotic	0,2573	1	0,0000	1	2,145	0,2573	ns
Herbaceous							
Native	2,8294	22	2,7679	18	1,969	1,6595	ns
Exotic	2,4129	10	2,2652	15	1,960	1,3656	ns
Lianas and others							
Native	2,2508	17	2,0208	14	1,960	1,2731	ns
Exotic	0,6792	4	1,0796	2	2,131	-0,1349	ns

* H' indices differ from each other ($P<0.05$)

ns- non-significant difference between H' ($P>0.05$)

Conventional crops had lower diversity indices of exotic species than agroecological ones, with a frequency four times higher. The diversity indices of native plant species were higher among the categories observed on the agroecological farms than on the conventional ones.

The high occurrence of herbaceous individuals among the conventional areas is due to the fact that there are fallow annual crop fields around the crops analysed, and shows that in conventional areas with a high degree of fragmentation, regeneration has a high rate of individuals of native species, with the possibility of regeneration through the seed bank that still exists. The exotic herbaceous individuals consist of spontaneous invasive plants in unused agricultural areas.

Most of the species considered lianas and others are bromeliads and lianas considered medicinal, such as timbó and são-joão (Pyrostegia venusta Miers), with other species such as caraguatá (Bromelia balansae), tree fern and species of açu, red and other ferns (Dicksonia spp.). In the surroundings of the agroecological crops, there was a good quality in the number of species of bromeliads, orchids, lianas and bamboos, with all four surroundings having more than 15 identified species. In the comparative conventional crops, only two neighbourhoods (out of a total of 5) had more than 12 species of these plants.

4.2 Occurrence of insect pests and diseases in experimental agroecological crops with high surrounding forest biodiversity and in comparative conventional crops with low surrounding forest biodiversity

The frequency of insect pests (Table 5), in general, as well as the diversity of species

(Figure 13) was higher in conventional crops.

Table 5 - Average number of insect pests in tomato crops in experimental agroecological and conventional comparative management systems in the municipality of Apiaí-SP. Mean ± SD (N=5)

	Cultivation	
	Experimental agroecological	Comparative conventional
Tomato moth (Tuta absoluta)	0,26 ± 0,19	18,00 ± 5,83
Whitefly (Benisia tabaci)	0,08 ± 0,06	34,00 ± 1.87
Little cow (Diabrotica speciosa)	0,61 ± 0,37	7,00 ± 2,00
Leaf miner (Liriomyza spp.)	0,00 ± 0,00	14,00 ± 4,00
Small tomato fruit borer (Neoleucinodes *elegantalis)* (% of fruit attacked/plant)	38,1 ± 5,14	33,00 ± 4,06

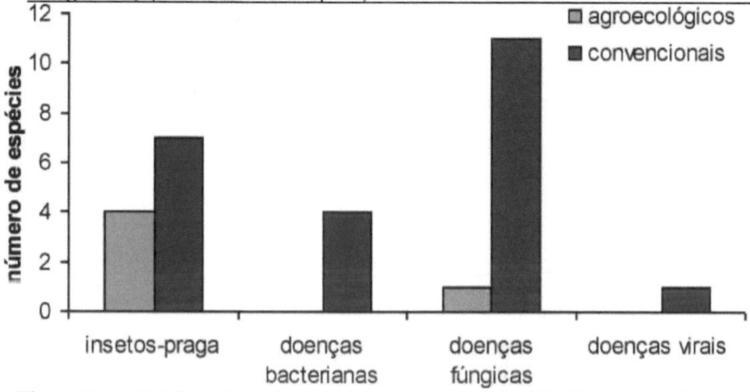

Figura 13 - Total number of pest and disease species identified in crops under experimental and conventional agroecological management compared

Conventional crops had a higher biodiversity index of insect pests (H=1.482) than agroecological crops (H=0.511), which in turn indicates a lower biodiversity of insect pests in agroecological crops.

Conventional tomato crops traditionally have a high population of insect pests (GRAVENA, 2003), whose occurrence is related to various factors such as the growing environment, the variety used, plant nutrition and the use of pesticides (CHABOUSSOU, 2006), among others (ALTIERI; NICHOLLS, 2003). However, the possibility of these communities establishing themselves in such a way as to cause agricultural damage is linked to many other factors, such as local crop management, the environment and the supply of nutrients accessible in the landscape where they are located (PANIZZI; PARRA, 1991).

In tomato plantations in places with high forest fragmentation, associated with the presence of various exotic annual crops favourable to the nutritional needs of insect pests, such as maize, pods, beans and various vegetables, the presence of "refuges" or alternative food sources for these insect communities should be considered, which helps to regulate populations and maintain communities of natural enemies throughout the year (LARA, 1979).

In studies carried out in the Azores region by Heleno et al. (2008), the replacement of the native forest component with exotic species can cause a reduction of up to 67 per cent in insect biomass, as well as a reduction in insects considered to be natural enemies of pest insects.

Bihn; Gebauer; Brandl (2010) demonstrate a close relationship between species and their functional diversity in Atlantic forest environments, where the forest component was gradually replaced by secondary and invasive species, which altered the population of ants that lived there naturally to species that adapt to the supply of new food and new ecological conditions (ODUM, 2007).

A characteristic found in agroecological areas with Atlantic forest is the constant presence of insects considered to be natural enemies of agricultural pests (ALTIERI, 2003), which, as cited by DeLoach (1970), occurs due to the maintenance of their natural habitats. Not only is the diversity and occurrence of parasitoids greater in crops surrounded by great forest diversity, but the rate of parasitism can also be favoured (MARINO; LANDIS, 1996).

During the field collections in forest environments, breeding and refuge sites were identified for various predator species of insect pests, such as wasps, birds, coleopterans, carnivorous ants, spiders and others, which access and interact in environments with borders made up predominantly of herbaceous and annual species due to the presence of insects with food sources. These natural predators have not been identified in the areas surrounding conventional tomato plantations, probably due to the intense application of agrochemical products and repellents of animal biodiversity (GRAVENA, 1992).

The edge of forests seems to harbour more species of generalist predators of the order Neuroptera than the interior, with the greatest diversity of species found in shrubs and trees (DUELLI; OBRIST; FLUCKIGER, 2002). The abundance of predators is also associated with plant diversity (LUNDGREN; WYCKHUYS; DESNEUX, 2009).

According to Gandara (2009), less intensive soil management associated with greater production of litter, such as in agroforestry systems, favours the diversity of macro invertebrates in the soil. As mentioned by Altieri; Nicholls (2003), natural predators play a role in regulating populations of phytophagous insects, which is a technical indicator for concluding that these agents are responsible for the low or no frequency of pest insects, which are generally quite common in the tomato plants sampled among agroecological producers, such as whitefly, thrips, the absence of miner and Bahia bugs, miner caterpillars, vaquinha and others.

The occurrence and severity of attacks by phytophagous insects on agricultural crops are greater the lower the ecological and environmental conditions, as well as the organic matter content, structural conditions and soil fertility, as mentioned by Altieri; Nicholls (2003). There is evidence of greater diversity and occurrence of parasitoids and predators in the arboreal strata surrounding crops than in the herbaceous strata (SPERBER et al., 2004; DYER; LANDIS, 1997), directly associated with plant diversity (SPERBER et al., 2004).

The occurrence of some tomato diseases, such as wilt (Ralstonia solanacearum) or other secondary bacteria, are said to be symptomatic of compacted soils with poor drainage, characteristics more easily achieved in soils with a higher clay content and low levels of organic matter than in sandier soils rich in organic matter (THORNTON et al., 2008), those typical of the municipality of Apiaí - SP.

According to Thornton et al. (2008), wetter soils have a weaker structure and are more susceptible to compaction than dry soils. In the case of tomato growing, the soils are constantly irrigated and are prepared for cultivation in conditions of high humidity. As mentioned by Ragassi (2009), penetration resistance above certain values is limiting to crop root growth, which vary slightly in the literature, such as 1.5Mpa (GRANT; LAFOND, 1993; STALHAM et al. 2007) and 2.0Mpa (SILVA; KAY; PERFECT, 1994; TAYLOR; ROBERSON; PARKER, 1994; TORMENTA; SILVA; LIBARDI, 1998).

Under unfavourable conditions of root penetration in soils, Stalham et al. (2007) point to the inefficient use of irrigation and synthetic nutrients, which leads to greater susceptibility of cultivated vegetables to the occurrence of insect pests and agricultural diseases in the tomato crop.

The measurements of the penetration capacity, up to 2.0Mpa, of the soils of the agroecological (n=25) and conventional (n=25) producers showed average values of 44.45 (± 4.89) cm for the conventional producers and over 90 cm deep for all the agroecological producers. These values for the depth of compacted layers are indicative of the formation of "plough feet", which leads to a reduction in the total volume of oxygen in the soil and in the root exploration area, which in many cases leads to the occurrence of fungal and bacterial diseases in tomato plants (RAGASSI, 2009).

The non-occurrence of wilt (Ralstonia solanacearum) and other soil fungi and bacteria in the agroecological crops (Table 6) can basically be attributed to the lack of compaction in the cultivated soils and their high drainage capacity, which is why the soil remains suppressive to these pathogens, as recommended by Tokeshi et al. (1997). There were no occurrences of viruses in the experimental agroecological crops, and in the comparative conventional crops the average number of affected plants/crop was 1.4 (± 0.24). It is worth noting that the conventional crops used systematic applications of insecticides and fungicides, while the agroecological crops only used Bordeaux mixture to control late blight.

Table 6 - Occurrence of bacterial and fungal diseases in tomato crops in agroecological systems experimental and conventional comparative in the Apiaí-SP region

	Crops affected (%)	
	Agroecological experiments	**Comparative conventions**
Bacterial diseases		
Bacterial canker (Corynebacterium michiganense)	0,00	100
Small bacterial spot (Pseudomonas tomato)	0,00	100
Hollow stem (Erwinia carotovora)	0,00	100
Fungal diseases		
Late blight (Phytophthora infestans)	100,00	100,00
Black spot (Alternaria solani and Macrosporium solani)	0,00	60,00
Septoria (Septoria lycopersici)	0,00	80,00
Stemphylium spot (Stemphylium solani)	0,00	10,00
Hard fruit rot (Phytophothora spp.)	0,00	0,00

Phoma rot (Phoma destructiva)	0,00	0,00
Powdery mildew (Oidium lycopersici)	0,00	60,00
Sclerotinia rot (Sclerotinia sclerotiorum)	0,00	100,00
Grey fruit mould (Botrytis cinerea)	0,00	80,00
Sclerotium rot (Sclerotium rolfsii)	0,00	20,00
Anthracnose (Glomerella singulata, Colletotrichum phomoides)	0,00	100,00
Fusarium wilt (Fusarium oxysporum f. lycopersici)	0,00	100
Verticillium wilt *(Verticillium albo-atrum)*	0,0	100

It is believed that the occurrence of soil-related diseases among conventional producers is, among other factors, related to the degree of soil compaction (MALAVOLTA JR., 2004), which results in poor water drainage through the soil and impedes root penetration.

Among conventional tomato growers, the ideal soils are those where the forest component was immediately removed for this crop, which have a high organic matter content and good structuring (MOREIRA; SIQUEIRA, 2006) and consequently reduce the possibility of diseases related to poor drainage, compaction and structural loss (RAGASSI, 2009).

As mentioned by Moreira; Siqueira (2006), organic matter is an essential part of maintaining all soil biota, and its incorporation at depth is mainly due to the roots of tree species and the associated microbiota.

According to Moço et al. (2005), Forest Biodiversity plays a fundamental role in the formation and quality of the litter deposited on soils, which with higher levels of biodiversity will lead to a greater diversity of nutrients and organic components, providing greater support for a more biodiverse fauna due to the greater supply of food in the soils, which, as mentioned by Tokeshi et al. (1997), the suppressive characteristics of soils to the occurrence of pathogens in agricultural crops, a fact that was noted among the experimental agroecological producers who had their crops in environments with better structured and drained soils than the conventional comparative ones.

Among the agroecological properties monitored in this project, no signs of soil compaction were identified, nor was the occurrence of diseases related to poor drainage and structuring conditions, probably due to the organic matter and structuring provided by the forest component in the area and around the crops.

With regard to late blight (Phytophtora infestans), the areas with agroecological management had 100% of the plants sampled diseased, with approximately 75% of the leaf area affected, while 70% of the plants in conventional cultivation had less than 25% of the leaf area affected, demonstrating the widespread occurrence of this fungus and the efficiency of fungicide control in conventional crops (Table 7).

Table 7 - Severity of infestation (% of leaf area affected) by Phytophtora infestans on tomato plants (%) in experimental agroecological and conventional comparative crops

	Very high >	High 75-	Average	Low 25-	Very low

	75 %	50%	50-25%	10%	<10%
			% plants		
Experimental agroecological crops 1	40,00	60,00	0,00	0,00	0,00
2	65,00	20,00	10,00	5,00	0,00
3	0,00	0,00	0,00	90,00	10,00
4	65,00	35,00	0,00	0,00	0,00
5	50,00	50,00	0,00	0,00	0,00
average	**44,00**	**33,00**	**2,00**	**19,00**	**2,00**
Comparative conventional crops					
1	0,00	0,00	0,00	8,00	92,00
2	5,00	5,00	20,00	50,00	20,00
3	0,00	0,00	6,67	13,33	80,00
4	0,00	0,00	7,69	15,38	76,92
5	0,00	0,00	0,00	55,00	45,00
average	**1,00**	**1,00**	**6,87**	**28,34**	**62,79**

From the second year onwards, the agro-ecological producers obtained good results in controlling late blight with the use of Bordeaux mixture, which meant that the tomato crop cycle could be completed with high yields. Fungal diseases (such as late blight) dispersed by the wind with great efficiency (FRY; MIZUBUTI, 1998), or also through the transport of infected material or other vector agents, can be spread over great distances, which is probably one of the main reasons for their widespread occurrence among conventional and agroecological producers, since forest biodiversity has no direct influence on their dispersal or incidence.

1.3 Analysing the relationship between forest biodiversity around tomato plantations and the occurrence of insect pests in the crop

In the linear correlation analysis, we adopted the parameter of 10 crops for the interpretation formulation. With covariance data of 0.7481 (t= 3.1891; gl=8; P= 0.012) for tree biodiversity and 0.7933 (t= 3.686; gl=8; P= 0.006) for plant biodiversity, we can say with 99% certainty that there is a correlation between native plant biodiversity around tomato crops, both tree and total, and the occurrence of insect pests. In other words, the occurrence of insect pests was higher in areas with low forest biodiversity (Figure 14), however, the statistical results are not responsible for stating that the relationships between these factors are of direct influence (TRIOLA, 1998).

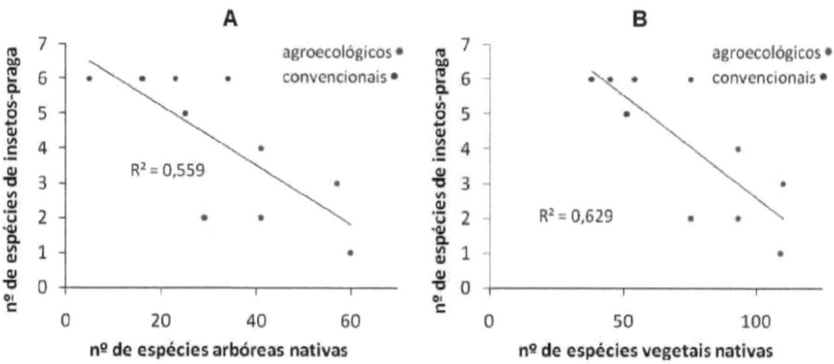

Figure 14 - Linear correlation between arboreal Forest Biodiversity (A) or plant biodiversity (B) and insect pest species in experimental and comparative conventional tomato crops in Apiaí - SP

Due to the similarity of climatic and environmental conditions between the different farms monitored, as well as the nutritional and health conditions of all the plants sampled, there are clear indications that the forest biodiversity around the agroecological crops has influenced the lower diversity and incidence of insect pests and pathogens compared to the conventional areas. Special attention should be paid to the diversity of native tree species, since it showed a significant difference in the indices between the agroecological and conventional crops (Table 4) and seems to be more associated with the diversity of natural enemies than the herbaceous strata (SPERBER et al., 2004; DYER; LANDIS, 1997).

1.4 Socio-environmental and economic analysis of tomato cultivation in agro-ecological and conventional management systems

In the Apiaí region, investments in conventional tomato crops vary according to the expected yield, from R$3.80/plant for estimated harvests of 200 boxes/thousand plants grown, to R$5.70 for rates above 300 boxes/thousand plants. As the conventional producers monitored by the project had average yields of around 250 boxes/thousand plants, their production costs were compatible with those expected for this production estimate, i.e. around R$4.26 per tomato plant

Among the experimental agroecological producers, there was a single cost of R$0.86 per foot cultivated, as all the inputs were purchased in one go and distributed equally among the producers over the two years of the project.

The total cost of producing a thousand tomato plants is five times higher among conventional producers than among agroecological producers (Figure 15), and the distribution of the cost of chemical pesticides among conventional producers is twice as high as the total cost of producing a thousand tomato plants in the agroecological model (Figure 16).

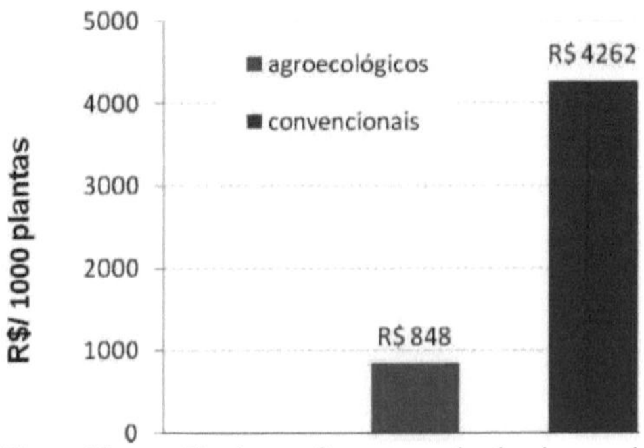

Figura 15 - Total cost of tomato production in experimental agroecological and conventional comparative crops in the Apiaí-SP region

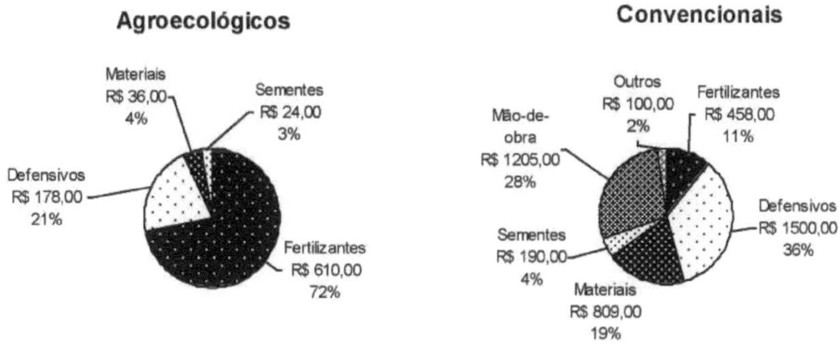

Figure 16 - Distribution of production costs for 1000 tomato plants in experimental agroecological and conventional comparative crops in the Apiaí-SP region

Of the total produced by the agroecological farmers, only part of the food produced was commercialised, with the priority of production being for internal consumption by the rural settlement community (Annex B). In order to estimate the profitability obtained by the agroecological producers, we used the parameters of total productivity, i.e. the volume marketed plus the volume of production consumed internally by the community.

The agro-ecological producers had tomato production costs on average five times lower than the average costs in the conventional model, and the maximum productivity obtained by the agro-ecological producers during the period of the experimental project was around 24% of the value obtained by the conventional producers, with productivity estimates among these producers, for this agroecological model, being around 40% of the average obtained among conventional producers, which would mean that planting 1000 tomato plants in an agroecological system on High Productivity Islands would show annual profitability of approximately 6.The average income obtained per box among agroecological producers was around R$68.75, 4.24 times higher than the average value obtained by conventional producers. Among the comparative conventional producers, no production losses were estimated, and the parameters of marketed productivity and gross profitability were taken into account for statistical analyses (Figure 17).

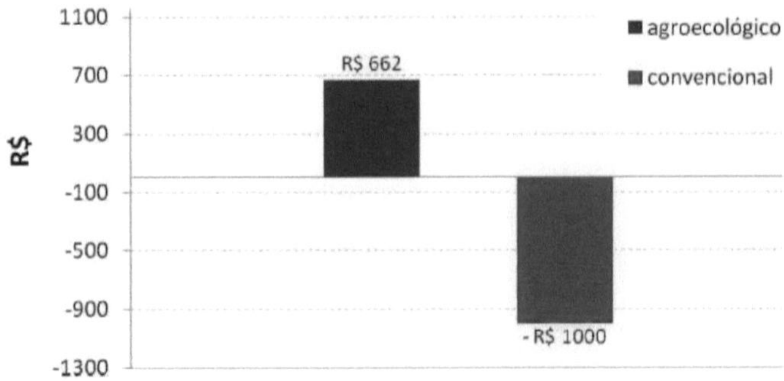

Figure 17 - Average net income from tomato cultivation in organic and conventional systems in Apiaí-SP

Of the five conventional producers monitored, two made a financial loss on their crops, and the other three conventional producers who made a profit on their activities did not make more than 7.8 per cent of the total amount invested, demonstrating, as mentioned by SANTOS; NORONHA (2001), that these are indicators of an economic activity that can show low profitability and has a high investment risk.

Among the conventional producers, we noticed that higher profitability rates are obtained from a strategy of seeking differentiated values in marketing, as in the case of producer C1, who obtained the highest profitability index per production (Appendix B), due to an individual marketing strategy and the search for differentiated market values.

Conventional producers are committed in their harvest to the agents who finance the activity and carry out the commercialisation, who determine the market price and payment conditions for the farmers.

Together with strategies to reduce the use of inputs and increase the value of productivity through the search for certified and differentiated foods, we can point to the need to build cooperative and community models of productivity and market access, as mentioned by Singer (2007); in order to expand the construction of a productive and social model focused on environmental conservation and the solidarity economy.

The producers involved in the management of the experimental modules between the two years of the project showed great appropriation of the agroecological management techniques disseminated for tomato cultivation.

As part of the discussion on productivity among agroecologists, we have to make it clear that the producers who overcame it by adopting successful strategies to control late blight had significant losses of around 50 boxes/1000 feet due to the occurrence of the small borer, which made it impossible to market and consume this part of the crop. In this way, more than 100 boxes/1000 feet were produced among the experimental models, such as the Zaqueu crop, which, based on the results in controlling the small borer, should reach this marketing potential, which projects approximate net profitability figures of 6000.00R$/1000 feet in the agroecological models in the High Productivity Islands.

The agroecological producers who took part in the project spent an average of 2 to 3 hours a day working, basically using the labour of the head of the family or, occasionally, the help of children and co-workers from other areas of the rural settlement in Apiaí - SP.

Conventional producers use hired labour, which accounts for an average of 28.27% of the cost of producing the crop, with employees having to dedicate ten to twelve hours of work a day to maintain a minimum of five thousand plants/person (Appendix C).

Conventional producers rent the land for tomato production and, because of this relationship with the land, they don't practise any strategy of consortium between cultivated species or any type of provision of environmental services, even using water for irrigation from springs and watercourses.

In all the agroecological farms, there were services added to tomato production, such as the intercropping of tomatoes with other plants of interest, such as cassava, corn or others, as well as the use of fruit unsuitable for consumption as animal feed.

Among the conventional producers, product losses were noted due to a variety of factors, ranging from the discarding of fruit attacked by borers and caterpillars to the loss of healthy fruit fit for consumption due to a lack of adequate prices that made harvesting and transporting the produce unfeasible.

Throughout this research, the agroecological producers stated that they had gained and increased their knowledge by taking part in a project to develop agricultural techniques in consortium with the Atlantic Forest, which demonstrates a social appreciation of their role as a farmer in developing agroecological techniques and valuing their cultural identity (LOPES; BERGAMASCO, 2009).

Although the financial reality of the two groups analysed is somewhat unstable in terms of their basic needs in terms of health, education, basic sanitation, transport and leisure, the conventional producers were more anxious and disbelieving about the possibilities of improvements in the agricultural environment with consequent social gains.

Conventional tomato growers had a high rate of investment in their crops, and interviews revealed that the agents who finance these crops are generally owners and businessmen in the food and vegetable trade in large urban centres such as Piracicaba and Campinas. These businessmen thus become responsible for transporting, commercialising and paying for these farmers' production, and when a loss occurs in a field, the balance is usually transferred to the next crop.

Most of the agroecological farmers' production was destined for local consumption and government food procurement programmes, and they should be more profitable if their productivity is certified as organic. Conventional producers often sold their harvest to the same person who financed the planting, demonstrating an isolated production and marketing strategy where production is destined for consumers in large urban centres such as São Paulo, Sorocaba and Campinas.

According to Gliessman (2001), there is a global need to seek sustainable production systems through the development of alternative agricultural practices, guided by knowledge of the ecological processes that occur in productive areas and in the wider environmental contexts of which they are part.

Peasant family farming plays an essential role in participating in and building these new production models (PETERSEN, 2009) and, according to Costa; Campanhola (1997), there have been growing numbers of producers seeking to operate under the organic and natural production system in the state of São Paulo.

The agroecological modules developed in the field with the participation of local farmers have an experimental purpose in terms of measuring the occurrence of insect pests and diseases. However, based on the improvements and development of technologies related to community management (BOEF et al., 2007), ecological pest management (PRIMAVESI, 1988) and the use of biodiversity in agricultural management (ALTIERI, 2003), the agroecological producers involved in the project predict productivity rates of around 100 boxes per thousand plants.

The participating agroecological producers are involved in activities related to the environmental recovery of springs and Permanent Preservation Areas (PPAs) and Legal Reserves (LRs), such as tree planting in joint efforts, among others, and in this way ensure compliance with environmental legislation which, according to Brancalion et al. (2010), are important instruments for the expansion and maintenance of natural biodiversity.

Due to the relatively low financial investment and profitability possibilities involved in this agro-ecological management proposal using Forest Biodiversity as an input for agricultural production, it can be developed and adapted to the needs and profile of family farmers and tomato producers in the Apiaí - SP region.

No planning or activities linked to environmental recovery were noted on the conventional properties, probably because the land used for production is usually owned by third parties.

The agro-ecological producers did not need irrigation to maintain the crop due to the planting season, which makes synchronisation with the seasonal climate of the local environment a strategy for reducing costs and inputs.

Agroecological crops required spraying with mineral and organic products, such as Bordeaux mixture and plant preparations (chilli and garlic), at a similar frequency to conventional producers, due to the high occurrence of the fungus that causes tomato late blight in this region (MINAMI, 1980) and the small tomato borer, which is responsible for a large part of the plant losses in these producers' crops.

A long history of incidents related to the health of farmers involved in tomato cultivation, resulting directly from the use of agrochemicals, such as fainting, dizziness and problems with gastric and liver function, was confirmed in the field through conversations and interviews with producers. According to Gravena (2003), and as found in the field research, tomato growing is conducted under intense application of agrochemicals, often combined, and in conventional crops the average was 36.4 sprays per cycle, higher than the national average of 30 sprays (GRAVENA, 2003).

A total of 29 chemical products were used in conventional crops, in a classification pattern similar to that cited by Santos; Noronha (2001), for table tomato crops in the Goiânia - GO region, with seven belonging to toxicological class I (Highly Toxic), four to class II (Very Toxic), eight to class II (Toxic) and ten to class IV (Slightly Toxic), all of which are highly dangerous to human health and the environment, according to Andrei (1996).

The use of personal protective equipment (PPE) has been noted among conventional tomato growers in the Apiaí region, but irregularly because they are often considered uncomfortable, as mentioned by Vicente et al. (1998), or with essential equipment missing such as gloves and others, so that one can consider the possibility of contamination and intoxication among the workers involved in agricultural management.

Among the agroecological producers, natural and mineral products such as limestone and sulphur were used, along with preparations of tobacco (Nicotiana tabacum), spring (Bougainvillea spectabilis), garlic (Allium sativum) and other forest products, such as cedar leaves (Cedrela fissilis), as well as insect baits such as tajujá (Cayaponia tayuya), a native liana, which attracts vines and other insects and disperses them from the crop of interest (SANCHES; ISHIMURA, 2001).

CHAPTER 5

CONCLUSIONS

1 - The incidence of insect pests in tomato crops was shown to be inversely correlated with the biodiversity present around the crops. Tomato crops in an agroecological management system, with forest biodiversity present in their surroundings, had a lower incidence of insects than conventional crops, even though the latter made severe use of repellent agrochemicals and insecticides.

2 - The occurrence of fungal and bacterial diseases originating in the soil was higher in the conventional crops than in the agroecological experiments. This is due to the fact that the conventional crops had a shallower soil compaction layer, which hinders drainage and favours the development of these diseases.

3 - The presence of forest biodiversity around the crops was not a limiting factor in the occurrence of the fungus Phytophthora infestans (late blight), which spreads through the air and is present in all the crops monitored in the area, the control of which was only possible with the use of organic inputs (organic farmers) or chemical fungicides (conventional farmers).

4 - The high economic and environmental cost of growing tomatoes in a conventional system is mainly due to the dependence on synthetic chemical inputs (fertilisers and agrochemicals), which increases the cost of installation and cultivation. The socio-environmental and health costs for conventional producers were high, due to the fact that they significantly reduced the conditions required for the health and well-being of producers and for the production of healthy food.

5 - It was concluded experimentally that forest biodiversity around tomato plantations can be an effective direct and indirect instrument for maintaining agricultural production and landscape balance, helping to regulate insect pest and disease populations. This issue is of great importance for family farmers with low purchasing power to obtain income and quality of life, using biodiversity as a tool. Biodiversity is therefore an input that, if used well, can reduce production costs and improve income, labour quality and the end product.

REFERENCES

ALTIERI, M.A. Developing sustainable agricultural systems for small farmers in Latin America. Natural Resources Forum, New York, v. 24, p. 97-105, 2000.

______. Agroecology: the science of natural resource management for poor farmers in marginal environments. Agriculture, Ecosystems and Environment, Amsterdam, v. 93, p. 1-24, 2002.

ALTIERI, M.A.; NICHOLLS, C.I. Soil Fertility management and insect pests: harmonising soil and plant health in agroecosystems. Soil & Tillage Research, Amsterdam, v. 72, p. 203-211, 2003.

ALTIERI, M.A.; SILVA, E.N.; NICHOLLS, C.I. O papel da biodiversidade no manejo de pragas. Ribeirao Preto: Editora Holos, 2003. 226 p.

ALTIERI, M.A.; TODD, J.W.; HAUSER, E.W.; PATTERSON, M.; BUCHANAN, G.A.; WALKER, R.H. Some effects of weed management and row spacing on insect abundance in soybean fields. Protection Ecology, Amsterdam, v. 3, n. 4, p. 339 - 343, 1981.

ALVARENGA, M.A.R. Botanical origin and description of plants. In: ALVARENGA, M.A.R. (Ed.). Tomatoes: field, greenhouse and hydroponic production. Lavras: Editora UFLA, 2004. p. 15-18.

ANDREI, O. Practical guide to phytosanitary products for agricultural use. 5. ed. São Paulo: Andrei Organisation, 1996. 506 p.

BRAZILIAN SEED AND SEEDLING TRADE ASSOCIATION. Tomatoes lead growth and profitability in the vegetable sector, 2010. Available at: <http://www.abcsem.com.br/noticia.php?cod=284>. Accessed on: 15 October 2010.

BARBIERI, R.L.; STUMPF, E.R.T. Origin and evolution of cultivated plants. Brasília: Embrapa Informação Tecnológica, 2008. 909p.

BETTIOL, W.; GHINI, R. Plant protection in alternative agricultural systems. In: CAMPANHOLA, C.; BETTIOL, W. Alternative control methods. Jaguariúna: Embrapa Meio Ambiente, 2003. p. 79-93.

BIHN, J.F.; GEBAUER, G.; BRANDL; R. Loss of functional diversity of ant assemblages in secondary tropical forests. Ecology. Washington, v. 91, n.3, p. 782-792, 2010.

BOEF, S.; THIJSSEN, M. H.; OGLIARI, J. B.; STHAPIT, B. R. Biodiversity and farmers; strengthening community management. Porto Alegre: LP&M, 2007. 271 p.

BRANCALION, P.H.S.; RODRIGUES, R.R.; GANDOLFI, S.; KAGEYAMA, P.Y., NAVE, A.G.; GANDARA, F.B.; BARBOSA, L.M.; TABARELLI, M. Legal instruments can enhance high-diversity tropical forest restoration. Revista Arvore, Viçosa, v. 34, n. 3, p. 455 - 470, 2010.

BRAZIL. Federal Decree 750/93, of 10 February 1993. Provides for the cutting, exploitation and suppression of primary vegetation, or vegetation in the advanced and medium stages of regeneration of the Atlantic Rainforest, and makes other provisions. Available at: <http://www.mp.ba.gov.br/atuacao/ceama/material/legislacoes/reserva/decreto_750_19 93.pdf.>. Accessed: 19 August 2010.

______. Federal Law n. 9.985 of 18 July 2000. Regulates article 225, § 1, items I, II, III and VII of the Federal Constitution, institutes the National System of Nature Conservation Units and makes other provisions. Available at: <http://www.planalto.gov.br/ccivil_03/Leis/L9985.htm>. Accessed: 19 August 2010.

______. Ministry of the Environment. National Environment Council. Resolution no. 10, of 01 October 2003. Published in the DOU: 03/11/1993.
Available at: <www.mma.gov.br/port/conama/res/res93/res1093.htm >. Accessed: 19 August 2010.

CHABOUSSOU, F. Plants diseased by the use of pesticides: new bases for prevention against diseases and parasites: the theory of trophobiosis. Translated by M. J.
Guazzelli. São Paulo: Expressão Popular, 2006. 320p.

CONSUEGRA, O.P.; MORALES, A.C.; LATERROT, H.E.; ANAÍS, G.B.J. Genetic improvement and management of tomato cultivation for production in the Caribbean. Cuba: Ministry of Agriculture, 2000. 319

p.

COSTA, M.B.B.; CAMPANHOLA, C.A Alternative agriculture in the state of São Paulo. Jaguariúna: Embrapa - CNPMA, 1997. 63 p.

DEAN, W. A ferro e fogo: a história e a devastação da Mata Atlântica Brasileira. São Paulo: Companhia das Letras, 1996. 484 p.

DeLOACH, C.J. The effect of habitat diversity on predation. Proceedings Tall Timber Conference on Ecological Animal Control by Habitat Management, Tallahasse, v. 2, p. 223-242, 1970.

DIEGUES, A.C. O mito moderno da natureza intocada. São Paulo: Editora Hucitec, 1996a. 169 p.

______. Human Populations in Protected Natural Areas of the Atlantic Forest. São Paulo: USP, NUPAUB. 1996b. Available at: <http://www.usp.br/nupaub/ConflitosnaMataAtlantica.pdf>. Accessed: 08 October 2010.

DUELLI, P.; OBRIST, M.K.; FLUCKIGER, P.F. Forest edges are biodiversity hotspots - Also for Neuroptera. In: 7th International Symposium on Neuropterology, 2000, Budapest. Acta Zoologica Academiae Scientiarum Hungaricae. Budapest, v. 48, suppl. 2, p. 75-87, 2002.

DYER, L.E.; LANDIS, D.A. Influence of noncrop habitats on the distribution of Eriborus terebrans (Hymenoptera: Ichneumonidae) in cornfields. Environmental Entomology. Lanham, v. 26, n.4, p. 924-932, 1997.

EHRLICH, P.R. The loss of biodiversity - causes and consequences. In: WILSON, E.O. (Ed.). Biodiversity. Rio de Janeiro: Nova Fronteira, 1997. p. 27-35.

EMBRAPA. National Research Centre for Vegetables. Growing tomatoes (Lycopersicon esculentum Mill). Brasilia, 1992. 22 p. (Document, 11).

FAGUNDES, A.F.; ONUKI, N.S.; RAUPP, D.S.; GARDINGO, J.R.; BORSATO, A.V. Influence of the degree of humidity on the texture of chilled or oil-filled dried tomatoes. Exact and Earth Sciences, Agricultural Sciences and Engineering, Ponta Grossa, v. 11, n. 1, p. 35-42, 2005.

FAO. State of the world's forests. Rome, 2007. p. 37-44,

FILGUEIRA, F.A.R. Manual de olericultura. São Paulo: Agronômica Ceres, 1972. 451 p.

FREIRE, P. Pedagogy of tolerance. Organisation and notes Ana Maria Araújo Freire. São Paulo: Editora UNESP, 2004. 329 p.

FRY, W.E.; MIZUBUTI, E.S. Potato late blight. In: JONES, D.G. (Ed.) The epidemiology of plant diseases. Dordrecht: Kluwer Academic, 1998. p. 371-388.

GALLO, D.; NAKANO, O.; SILVEIRA NETO, S.; CARVALHO, R.P.L.; BAPTISTA, G.C.; BERTI FILHO, E.; PARRA, J.R.P.; ZUCCHI, R.A.; ALVES, S.B.; VENDRAMIM, J.D.; MARCHINI, L.C.; LOPES, J.R.S.; OMOTO, C. Entomologia agrícola. Piracicaba: FEALQ, 2002. 920 p.

GANDARA, F.B. How much biodiversity can agroforestry systems maintain? case study of spontaneous plants and soil macro invertebrates in Piracicaba, SP. In: LATIN AMERICAN ECOLOGY CONGRESS, 3., 2009, São Lourenço. Proceedings... p.1-3

GLIESSMAN, S.R. Agroecology: ecological processes in sustainable agriculture. 2. ed. Porto Alegre: Editora Universidade UFRGS, 2001. 653 p.

GRANT, C.A.; LAFOND, G.P. The effects of tillage systems and crop sequences on soil bulk density and penetration resistance on a Clay soil in Southern Saskatchewan.
Canadian Journal Soil Science. Ottawa, v. 73, n. 2, p. 223-232, 1993.

GRAVENA, S. Biological control in integrated pest management. Pesquisa Agropecuária Brasileira, Brasília, v. 27, p. 281-299, 1992.

GRAVENA, S.; BENVEGA, R. S. Practical manual for the ecological management of tomato pests. Jaboticabal: Gravena-ManEcol, 2003. 143p.

GRAVENA, S.; CHURATA MASCA, M. G. G.; ARAI, J.; RAGA, A. Integrated Management of the Whitefly - Bemisia tabaci (Gennadius, 1889) in cultivars of determinate tomato with a view to reducing golden mosaic virus. Anais da Sociedade Entomológica do Brasil, Londrina, v. 13, n. 1, p. 35-45, 1984.

GRELLE, C.A.V.; PINTO, M.P.; MONTEIRO, J.; FIGUEIREDO, M. S. L. A decade of Conservation Biology in Brazil. Oecologia Brasiliensis, Rio de Janeiro, v. 13, n. 3, p. 420-433, 2009.

HARPER, J.; HAWKSWORTH, D. Preface. In: HAWKSWORTH, D.L. (Ed.) Biodiversity, measurement and estimation. London: Chapman Hall, 1995. p. 5-11.

HELENO, R.H.; CEIA, R.S.; JAIME, J.A.; MEMMOT, J. Effects of alien plants on Insect abundance and biomass: a food-web aproach. Conservation Biology, Oxon, v. 23, n.2, p. 410 - 419, 2008.

HEYER, W.R.; RAND, A.S.; CRUZ, C.A.G.; PEIXOTO, O.L. Decimations, extinctions and colonisations of frogs populations in Southest Brazil and their evolutionary implications. Biotropica, Washington, v. 20, p. 230-235, 1988.

BRAZILIAN INSTITUTE OF GEOGRAPHY AND STATISTICS. Cities - Apiaí - SP.
Population and households - Census 2000 with territorial division 2001. 2001. Available at: <http://www.ibge.gov.br/cidadesat/topwindow.htm?1>. Accessed: 19 August 2010.

______. Estimates of resident populations on 1 July 2009, by municipality. 2009. Available at: <http://www.ibge.gov.br/home/estatistica/populacao/estimativa2009/POP2009_DOU.pdf >. Accessed: 19 August 2010.

______. 2009. Situation of Brazilian Vegetable Production - 2008. Available at: <http://www.cnph.embrapa.br/paginas/hortalicas_em_numeros/hortalicas_em_numeros. htm>. Accessed: 15 October 2010.

JAMES, W.C. An illustrated series of assessment keys for plant diseases. Their preparation and usage. Cannadian Plant Disease Survey, Quebec, v. 51, n. 2, p. 3965, 1971.

JEFFRIES, M. Biodiversity and conservation. London: Routledge, 1997.

KAGEYAMA, P.Y.; GANDARA, F.B.; SOUZA, L.M.I. Consequencias genéticas da fragmentação sobre populações de espécies arbóreas. Série Técnica IPEF, Piracicaba, v. 12, n. 32, p. 65-70, 1998.

KAGEYAMA, P.Y.; OLIVEIRA, R. S.; FERRAZ, P. A.; FURTADO, E. L.; SOUZA, A. D.; SEBBENN, A. M. Genetic gains in the selection for latex productivity in natural populations of Hevea brasiliensis in Chico Mendes reserve. A case study of the IAPs (Islands of High Productivity). Scientia Forestalis/Forest Sciences, Piracicaba, v. 61, p. 79-85, 2002.

KUROZAWA, C.; PAVAN, M.A. Tomato diseases. In: KIMATI, H.; AMORIM, L.; BERGAMIN FILHO, A.; CAMARGO, L.E.A.; REZENDE, J.A.M. (Ed.). Manual de fitopatologia: doenças das plantas cultivadas. São Paulo: Agronômica Ceres, 2005. v. 2, p. 607-626.

LARA, F.M. Principles of plant resistance to insects. Piracicaba: Livroceres. 1979. 207 p.

LOPES, A.W.P.; BERGAMASCO, S.M.P.P. Agricultural Practices in Rural Settlements in the State of São Paulo: conventional and alternative forms of agroecological production. In: SYMPOSIUM ON AGRICULTURAL REFORM AND RURAL SETTLEMENTS, 4, 2010. Rural settlements 2010: controversies and development alternatives, proceedings... In: CD-ROM.

LOVEJOY, T. E. Changes in biological diversity. In: BARNEY, G.O. (Ed.) The global report to the president. Penguin Books, 1980. v. 2.: Harmondsworth, p. 327-332.

LUNDGREN, J.G.; WYCKHUYS, K.A.G.; DESNEUX, N. Population responses by Orius *insidiosus* to vegetational diversity. BioControl, Dordrecht, v. 54, p. 135-142, 2009.

LUTZEMBERGER, J. Plants sickened by the use of pesticides - The theory of trophobiosis. Agroecologia Hoje, Botucatu, year 3, n.16, p. 5-7, 2002.

MALAVOLTA JR., V. A. Bacterial diseases in tomato: etiology and control. Campinas: Agronomic Institute of Campinas. 2004. Tomato Production Chain Update Seminar 31/03/2004 and 01/04/2004. Mogi-Guaçu, SP. Available at <http://www.feagri.unicamp.br/tomates/pdfs/doebacter.pdf.> Access: 23 September 2010.

MARINO, P.C.; LANDIS, D.A. Effect of landscape structure on parasitoid diversity and parasitism in agroecosystems. Ecological Applications. Washington, v. 6, n. 1, p. 276284, 1996.

MATTOS, C.R.R.; GARCIA, D.; PINARD, F.; LE GUEN, V. Variability of Microcyclus uley isolates in southern Bahia. Fitopatologia Brasileira, Jaboticabal, v. 28, p. 502507, 2003.

MINAMI, K. The tomato plant. Campinas: Cargill Foundation, 1980. 397 p.

MOÇO, M.K.S.; GAMARODRIGUES, E.F.; GAMARODRIGUES, A.C.; CORREIA, M.E. Characterisation of edaphic fauna in different vegetation covers in the Northern Fluminense Region. Revista Brasileira de Ciências do Solo, Campinas, v. 29, p. 555-564, 2005.

MOREIRA, F.M.S.; SIQUEIRA, J.O. Microbiologia e bioquímica do solo. 2. ed. Lavras: Editora da Universidade Federal de Lavras, 2006. p. 203- 261.

ODUM, E.P. Fundamentals of ecology. 16. ed. São Paulo: Cengage Learning, 2007. 612 p.

OOSTRUM, J. Van. Networking: concept or practice? A participatory study with the Organic Matter Management Network members in Kakamega, Busia and Vihiga Districts. Nairobi: Association for Better Land Husbandry, 1996. 34 p.

OOSTRUM, J. Van, STIVEN, R. A strategy for the Association of Better Land Husbandry. Nairobi: Association for Better Land Husbandry, 1997. 34 p. (Report, 4).

PANIZZI, A.R.; PARRA, J.R. Ecologia nutricional de insetos e suas implicações no manejo de pragas. São Paulo: Manole, 1991. 359 p.

PETERSEN, P. Peasant family farming in the construction of the future. Rio de Janeiro: ASPTA, 2009. 168 p.

PIERRO, A. Good taste. Cultivar: Hortaliças e Frutas, Pelotas, v. 2, p. 10-12, 2002.

CITY HALL OF APIAI. The city of Apiaí. 2009. Available at: <http://apiai.sp.gov.br/cidade/index_.php?page=dados>. Accessed: 19 August 2010.

PRIMAVESI, A. Manejo ecológico de pragas e doenças. São Paulo: Nobel, 1988. 137 p.

______. Sustainable agriculture: manual for rural producers. São Paulo: Nobel, 1992. 142 p.

RAGASSI, C.F. Deep tillage system and grass succession for planting potatoes (Solanum tuberosum L.). 2009. 82 p. Dissertation (Master's Degree in Plant Science) - Escola Superior de Agricultura "Luiz de Queiroz", Universidade de São Paulo, Piracicaba, 2009.

ATLANTIC FOREST BIOSPHERE RESERVE -RBMA. Atlantic Forest Yearbook, 500 years of destruction. Available at: <http://www.rbma.org.br/anuario/mata_03_anosdedesttuicao.asp.> Accessed: 19 Aug. 2010.

RICK, C.M. Tomato. In: SIMMONDS, N.W. (Ed.) Evolution of crop plants, London: Longman, 1976. p. 269-273.

______. Tomato - Lycopersicon esculentum (Solanaceae). In: SMART, J.; SIMMONDS, N. W. Evolution of crop plants. London: Longman, 1995. p. 449-451.

SAMUELSON, A.P.; NORDHAUS, D.W. Cost analysis. In: SAMUELSON, A.P. (Ed.). Economics. 18. ed. Lisbon: McGraw-Hill. 1993. p. 139-181.

SANCHES, M.A.; ISHIMURA, I. Attractiveness of taiuiá seeds to Diabrotica speciosa in chard. Instituto Biológico, São Paulo, v. 68, n. 2, p. 97-101, 2001.

SANTOS, J.D.; KAGEYAMA, P.Y.; GUSSON, E.; CUNHA, G. Diagnosis of the vegetation in the forest areas of the Luiz David de Macedo settlement - Apiaí/SP. Research Programme for the Development of Rural Settlements and Family Farming, Piracicaba: INCRA; FEALQ, 2008. 82 p.

SANTOS, M. M.; NORONHA, J. F.; Diagnosis of table tomato cultivation in the municipality of Goianópolis, State of Goiás - Brazil. Pesquisa Agropecuária Tropical, Goiânia, v. 31, p. 29-34. 2001

SÃO PAULO (State). Department of the Environment. Forestry Foundation. Conservation Units - State Parks. Available at: < http://www.fflorestal.sp.gov.br/parquesEstaduais.php>. Accessed: 28 October 2010.

SHANNON, C.E.; WEAVER, W. The mathematical theory of communication. Urbana: University of Illinois Press, 1949. 117 p.

SHIMADA, H. Mina do Morro do Ouro, Apiaí (SP): from environmental liability to park. In: BRAZILIAN CONGRESS OF GEOLOGY, 44, 2008, Curitiba. 1 CD-ROM.

SILVA, A.P.; KAY, B.D.; PERFECT, E. Characterisation of the least limiting water range of soils. Soil Science Society of America Journal. Madison, v. 58, p. 1775-1781, 1994.

SILVA, J.B.C.; LOPES, C.A.; MAGALHÃES, J.S. Sweet potato cultivation. Production Systems 6. Electronic version. 2004. Embrapa Hortaliças. Available at: <http://www.cnph.embrapa.br/sistprod/batatadoce/index.htm>. Accessed: 08 October 2010.

SILVA, J.B.C. da; GIORDANO, L.B.; FURUMOTO, O. BOITEUX, L.S.; FRANÇA, F.H.; VILLAS BÔAS, G.L.; CASTELO BRANCO, M.; MEDEIROS, M.A.; MARQUELLI, W; SILVA, W.L.C. e; LOPES, C.A.; ÁVILA, A.C. ; NASCIMENTO, W.M.; PEREIRA, W. Growing tomatoes for industrialisation. Brasília: Embrapa Hortaliças, 2003. Production Systems, 1. Available at: <http://sistemasdeproducao.cnptia.embrapa.br/FontesHTML/Tomate/TomateIndustrial_ 2ed/index.htm>. Accessed: 20 September 2010.

SILVA, M.J.P. da. Reca: the basis of alternative agriculture with economic efficiency. Revista Brasileira de Agroecologia, Porto Alegre, v. 1, n. 1, p. 1259-1262, 2006.

SINGER, P. Solidarity economy: democracy and conflicts between equals. Otra Ecomonia, v. 1, n. 1, p. 14-16, 2007. Available at: < http://www.riless.org/otraeconomia/singer.pdf>. Accessed: 25 October 2010.

SMITH, C.M.; BRUMMIT, N.A.; OLIVEIRA-FILHO, A.T.; BACHMAN, S.; MOAT, J.; NICH LUGHADHA, E.M.; LUCAS, E.J. Plant diversity hotspots in the Atlantic Coastal Forests of Brazil. Conservation Biology, Boston, v. 23, n. 1, p. 151-163, 2008.

SOS MATA ATLÂNTICA. Information about the Atlantic Forest. Available at: <http://www.sosmatatlantica.org.br/index.php?section=info&action=flora>. Accessed: 19 August 2010.

SPERBER, C.F.; NAKAYAMA, K.; VALVERDE, M.J.; NEVES, F.S. Tree species richness and density affect parasitoid diversity in cacao agroforestry. Basic and Applied Ecology, Jena, v. 5, p. 241-251, 2004.

STALHAM, M.A.; ALLEN, E.J.; ROSENFELD, A.B.; HERRY, F.X. Efects of soil compaction in potato (Solanum tuberosum L.) crops. Journal of Agricultural Science, Cambridge, v. 145, p. 295-312, 2007.

TAVARES, C.A.M. Virus attack. Cultivar: Hortaliças e Frutas, Pelotas, v. 2, n. 17, p. 26-29, 2003.

TAYLOR, H.M.; ROBERSON, G.M.; PARKER, J.J. Soil strength-root penetration relations towards a system with multiple signalling. Soil and Tillage Research. Amsterdam, v. 30, p.217-243, 1994.

THORNTON, M.; STARK, J.; HOPKINS, B. G.; THORNTON, R. E. Selecting and preparing the planting site. In: JOHNSON, D. A. (Ed.). Potato health management. 2nd ed. Saint Paul: The American Phytopathological Society, 2008. chap. 5 p. 23-30.

TOKESHI, H.; ALVES, M.C.; SANCHES, A.B.; HARADA, D.Y. Control of Sclerotinia sclerotiorum with effective microorganisms. Summa Phytopathologica, Jaboticabal, v. 23, p.146-154, 1997.

TOMAS, F.L.; SALUSTIO, P.E.B.; KAGEYAMA, P.Y.; TOKESHI, H. Influence of surrounding forest biodiversity and distance between tomato (Lycopersicum sculentum) crops on the dynamics and occurrence of insects and diseases. Revista Brasileira de Agroecologia, Porto Alegre, v. 4, n. 2, p. 2425-2428, 2009.

TORMENTA, C.A.; SILVA, A.P.; LIBARDI, P.L. Characterisation of the optimum water interval of a purple latosol under no-tillage. Brazilian Journal of Soil Science. Viçosa, v. 22, p. 573-581, 1998.

TRIOLA, F.M. Introdução a estatística. 7. ed. Rio de Janeiro: LTC, 1998. p. 234-271.

VELOSO, H.P. Phytogeographic system. In: BRAZILIAN INSTITUTE OF GEOGRAPHY AND STATISTICS. Technical manual of Brazilian vegetation. Rio de Janeiro, 1992. p. 938. (Technical Manuals in Geosciences, 1).

VICENTE, M.C.M.; BAPTISTELA, C.S.L.; COELHO, J.P.; LOPES JUNIOR, A. Perfil do aplicador de agrotóxicos na agricultura Paulista. Informações Econômicas, São Paulo, v. 28, p. 35-47, 1998.

VIVAN, J.L. Pomar ou floresta: princípios para manejo de agroecossistemas. Rio de Janeiro: AS-PTA, 1993. 96 p.

______. Agriculture and forests: principles of a vital interaction. Guaíba: Agropecuária, 1998. 207 p.

WILSON, E.O. The current state of biological diversity. In: WILSON, E.O. (Ed.). Biodiversity. Rio de Janeiro: Nova Fronteira, 1997. p. 3-24.

ZAR, J.H. Biostatistical analysis. 5th ed. New Jersey: Pearson Prentice Hall, 2010. 944 p.

ANNEXES

Annex A- General data on the municipality of Apiai - SP

Data from the Municipality of Apiaí-SP	
Foundation	14th August 1771
Municipal area	1,005.73 km^2
Centre marker 1	Mother Church
Altitude	1050 m
Latitude	24°30'35.19311 South
Longitude	48°50'33.47928 West
Coordinated	SAD 69
Weather	Subtropical, Temperate/Super Humid
Topography	Mountainous with slopes and plateaus
Hydrography	Ribeira de Iguape, Pilões, Macacos, Tijuco, Palmital, Claro, Saivá and Estiva
Main soils	Massapés and Latosols
Population	27,136 inhabitants (IBGE, 2000)
Population density	26.98 inhabitants/km^2

Source: Prefeitura Municipal de Apiai - SP (2009)

Annex B Cost, productivity and profitability data for tomato producers in experimental agroecological and conventional comparative cultivation systems in the municipality of Apiaí-SP

	Cultivation systems	
	Agroecological	**Conventional**
Property 1		
Total number of boxes	18 commercialised (23 produced)	8.500
Cost/cash	R$ 37,00	R$ 22,35
Total cost of production	R$ 848,00	R$ 190.000,00
Gross income	R$ 1.125,00	R$ 205.000,00
Net income	R$ 277,00	R$ 15.000,00
Average price/box	R$ 62,50	R$ 24,11
Property 2		
Total number of boxes	9 commercialised (13 produced)	5.000
Cost/cash	R$ 65	R$ 17,00
Total cost of production	R$ 848,00	R$ 185.000,00
Gross income	R$ 562,00	R$ 71.000,00
Net income	- R$ 285,50	-R$ 14.000,00
Average price/box	R$ 62,50	R$ 14,20
Property 3		
Total number of boxes	40 commercialised (55 produced)	10.640
Cost/cash	R$ 15	R$ 15,04
Total cost of production	R$ 848,00	R$ 160.000,00
Gross income	R$ 3.000,00	R$ 175.000,00
Net income	R$ 2152,00	R$ 15.000,00
Average price/box	R$ 75,00	R$ 16,45
Property 4		

Total number of boxes	3 commercialised (6 produced)	3.320
Cost/cash	R$ 141	R$ 13,16
Total cost of production	R$ 848,00	R$ 70.000,00
Gross income	R$ 225,00	R$ 76.000,00
Net income	- R$ 623,00	R$ 6.000,00
Average price/box	R$ 75,00	R$ 14,28
Property 5		
Total number of boxes	4 commercialised (8 produced)	4.050
Cost/cash	R$ 106	R$ 18,68
Total cost of production	R$ 848,00	R$ 75.650,00
Gross income	R$ 225,00	R$ 48.650,00
Net income	- R$ 623,00	- R$ 27.000,00
Average price/box	R$ 75,00	R$ 12,01

Annex C - Socio-economic characteristics of tomato producers in the municipality of Apiai - SP monitored by the project

Activity characteristic	Agroecological producer	Conventional producer
	2 producers in 2009/ 3 in 2010 PNRA settler	2 producers in 2009/ 3 in 2010 4 tenants, 1 owner
No. of producers accompanied and interviewed Relationship with the land they plant on Cost of land for planting Recipients of production	Own land Direct purchase/PAA projects Own funds, loan, project aid 2 to 3 h/day Own and family	R$ 1600.00/ ha (average) CEASA stones 3 loan of CEASA stone; 2 own costs 10 to 12 hours/day Employee - CLT
How production is costed Daily working time spent by the producer Type of labour employed Number of people involved in culture	1 person per thousand feet 36 combined sprays of organic products and neem Mineral and organic	1 worker for every 5,000 feet 36 combined puverisations (average) Synthetic and chemical
Average number of pesticide applications Nature of the inputs used Prospective use of the agricultural area (cycle of use)	One year with 3 years of fallow Approx. 3 months Not used	One year with at least 20 years of fallow Aluminium piping % and 1 in.
Harvest time and total production /1000 ft Irrigation equipment used Irrigation inputs (fuel, parts, other) Spraying equipment used Spraying inputs (nozzles, clothing, other) Transport equipment used History of contamination	Not used 20 litre backpack pump No irrigation Animal traction, tractors for community use None	Hydraulic pumps General maintenance of machinery Gas hydraulic tank with hoses and nozzles One PPE kit/harvest/worker Own tractors with trailers Recurrent dizziness and reports of malaise Complete for high-capacity pesticides. Contam Unidentified
Use of PPE	Waistcoats and masks for applying Bordeaux mixture	
Environmental recovery or compensation activities	Averaging, reforestation, and APP and RL maintenance	

MIX
Papier aus verantwortungsvollen Quellen
Paper from responsible sources
FSC® C105338

Printed by Books on Demand GmbH, Norderstedt / Germany